Lecture Notes in Intelligent Transportation and Infrastructure

The series "Lecture Notes in Intelligent Transportation and Infrastructure" (LNITI) publishes new developments and advances in the various areas of intelligent transportation and infrastructure. Merging theoretical foundations, practical applications, and forward-looking insights, LNITI provides a comprehensive understanding of both the state-of-the-art and the future prospects within this dynamic field.

LNITI is designed to be an inclusive platform that covers an extensive array of topics including, but not limited to intelligent transportation systems, smart mobility, intelligent logistics, critical infrastructure, smart architecture, smart cities, intelligent governance, construction design, data security, operational analysis, optimal route planning, digitalization, autonomous vehicles, the evolution of transport systems as well as green and sustainable urban structures. The series contains monographs, conference proceedings, edited volumes, lecture notes and textbooks. Of particular value to both the contributors and the readership are the short publication timeframe and the world-wide distribution, which enable wide and rapid dissemination of high-quality research output.

Proceedings published in the series are indexed by INSPEC.

All books, including proceedings, published in the series are evaluated by Web of Science.

Elisa Mariavittoria Bertolini

Temporal Value in MCDM

Enhancing Infrastructure Planning and Connectivity

Springer

Elisa Mariavittoria Bertolini
Department of Engineering
and Architecture
University of Trieste
Trieste, Italy

ISSN 2523-3440 ISSN 2523-3459 (electronic)
Lecture Notes in Intelligent Transportation and Infrastructure
ISBN 978-3-031-95484-9 ISBN 978-3-031-95485-6 (eBook)
https://doi.org/10.1007/978-3-031-95485-6

This Springer imprint is published by the registered company Springer Nature Switzerland AG
The registered company address is: Gewerbestrasse 11, 6330 Cham, Switzerland

Preface

The planning and development of transport infrastructure represent one of the most consequential areas of public policy, directly influencing territorial organization, economic development, environmental sustainability, and social equity. As urban and regional systems become increasingly complex and interdependent, the need for innovative and adaptable planning methodologies becomes ever more urgent. In this context, transport infrastructure is not simply a matter of engineering and logistics—it is a deeply strategic function of governance, negotiation, and long-term vision.

This monograph emerges from a desire to advance the tools and methods used in infrastructure planning by introducing a framework that incorporates temporal value, stakeholder dynamics, and group decision-making principles. Traditional evaluation models—most notably, cost-benefit analysis—often fall short in capturing the nuanced, long-range impacts of infrastructure investments, particularly when it comes to social equity, environmental sustainability, and political feasibility. This work, therefore, proposes an integrated and participatory approach grounded in multi-criteria decision analysis (MCA), one that explicitly incorporates time as a central variable and acknowledges the complexities of strategic interactions among stakeholders.

The methodology developed here provides an innovative framework that not only quantifies stakeholder preferences over a 30 year horizon but also models group decision-making behavior, power asymmetries, and the emergence of strategic alliances. It introduces novel mechanisms for capturing temporal value in long-term projects and demonstrates how such evaluations can be embedded in participatory governance structures to enhance both legitimacy and effectiveness in public infrastructure planning.

This monograph is the result of research carried out within the framework of transport and logistics planning, with the Gorizia Freight Terminal (Interporto) serving as the empirical foundation. This strategic infrastructure, located in the Friuli Venezia Giulia region, is emblematic of the planning challenges faced by cross-border territories seeking to integrate sustainability, competitiveness, and multi-modal connectivity. The case study allows for the application and testing of the proposed methodology in a real-world context, offering valuable insights into how policy frameworks can be adapted to support more inclusive and forward-looking infrastructure decisions.

The text is structured into 12 chapters, each addressing a specific dimension of the planning and evaluation process.

Chapter 1 introduces the general context and core research questions.

Chapter 2 discusses strategic territorial choices, drawing attention to the complexities of decision-making in regional development.

Chapter 3 reviews decision-support methodologies, particularly MCA, within the transport and infrastructure literature.

Chapter 4 deepens the discussion on stakeholder participation and governance through the lens of group decision-making.

Chapter 5 presents the methodological framework, which integrates multi-criteria analysis with a temporal perspective.

Chapter 6 introduces the concept of the Working Group as a decision-making body that mediates across multiple levels of governance.

Chapter 7 focuses on the individual ranking of alternatives, proposing tools like the Simos method to derive stakeholder weights.

Chapter 8 addresses the evaluation of temporal value, offering a structured approach for considering long-, medium-, and short-term impacts.

Chapter 9 explains the adapted use of the MACTOR method to analyze stakeholder dynamics, strategies, and influence structures.

Chapter 10 applies the methodology to the Gorizia Freight Terminal, evaluating multiple development alternatives.

Chapter 11 elaborates on the performance analysis of each alternative, integrating stakeholder feedback and intertemporal comparisons.

Chapter 12 draws conclusions and discusses the broader implications of the research for infrastructure planning and territorial governance.

Trieste, Italy
April 2025

Elisa Mariavittoria Bertolini

Competing Interests The author has no competing interests to declare that are relevant to the content of this manuscript.

Contents

Chapter 1
Introduction

Abstract The development of territorial spaces, particularly through the lens of transport infrastructure planning, represents one of the most intricate and multifaceted challenges in contemporary public policy and urban development. Decisions made in this domain not only shape the spatial organization of regions but also significantly influence economic growth, social cohesion, and environmental sustainability. As such, transport infrastructure planning necessitates a comprehensive and dynamic approach that accounts for the interdependencies between technical, economic, and societal factors.

The development of territorial spaces, particularly through the lens of transport infrastructure planning, represents one of the most complex and multifaceted challenges in contemporary public policy and urban development. Decisions made in this domain not only shape the spatial organization of regions but also have a significant impact on economic growth, social cohesion, and environmental sustainability. Therefore, transport infrastructure planning requires a comprehensive and dynamic approach that considers the interdependencies between technical, economic, and societal factors.

Transportation systems play a pivotal role in territorial transformation, acting as a catalysts for socio-economic development beyond their function as physical networks of connectivity. Research suggests that transport infrastructure affects life quality and environmental outcomes by between 30 and 50%. The complexity of transport system planning goes far beyond engineering specifications, requiring the integration of public administration competencies, private property rights, societal welfare considerations, and economic sustainability. This interplay highlights the necessity of decision-making frameworks capable of addressing these diverse and often competing interests.

A key challenge in transport infrastructure planning is the inherent uncertainty of long-term decision-making. Forecasting future mobility patterns, economic trends, and technological advancements is complex and creates significant obstacles in the

E. M. Bertolini, *Temporal Value in MCDM*, Lecture Notes in Intelligent Transportation and Infrastructure, https://doi.org/10.1007/978-3-031-95485-6_1

planning process. Additionally, political considerations often influence infrastructure development, sometimes leading to the prioritization of projects based on short-term electoral objectives rather than long-term strategic value. Unlike simplified theoretical models, real-world decision-making processes are dynamic and require the engagement of multiple stakeholders with potentially conflicting interests. These processes must also navigate intricate legislative and regulatory frameworks, balancing strategic, long-term visions with immediate operational needs.

Given these challenges, transport infrastructure decision-making requires an advanced framework that considers economic sustainability, social impact, and environmental consequences. Traditional evaluation methodologies, such as cost–benefit analysis (CBA), have long been used to assess the feasibility of infrastructure projects by quantifying the economic benefits against the costs. However, critics argue that CBA often fails to capture non-monetary impacts such as social equity and environmental sustainability. To overcome these limitations, alternative approaches, such as multi-criteria decision analysis (MCA), incorporate qualitative factors alongside economic metrics. Additionally, scenario planning has gained prominence as a tool for addressing uncertainty by modelling potential future developments and their implications for infrastructure projects.

This monograph introduces an innovative methodological framework designed to address the multifaceted challenges inherent in infrastructure planning. The research recognises that effective infrastructure development requires more than technical expertise; a holistic and adaptive approach is needed to navigate complexity, balance diverse interests, and anticipate long-term implications. The proposed framework places temporal value at the centre of multi-criteria decision analysis. Through the developement of 30-year evaluation model, the study systematically assesses the performance of different infrastructure alternatives over extended periods.

Unlike conventional evaluation techniques, which often treat time as a peripheral consideration, the proposed methodology integrates temporal preferences directly into the multi-criteria evaluation process. By systematically eliciting and quantifying temporal value, this research provides decision-makers with a nuanced understanding of how infrastructure investments evolve and generate value over time. This methodological innovation provides practical benefits, including enhanced transparency in decision-making, a more comprehensive understanding of the long-term impacts of project, and improved alignment between strategic vision and operational implementation.

This research is based on empirical evidence from the Gorizia Interporto, which is used as a critical case study to test the proposed methodology. By applying the temporal multi-criteria evaluation approach to this specific infrastructure project, the study demonstrates the practical applicability and transformative potential of the framework. This novel framework provides a systematic approach to infrastructure planning, offering an in-depth analysis of stakeholder behaviour in group decision-making contexts and incorporating temporal preferences in multi-criteria evaluations. It also develops a mechanism for understanding power dynamics and potential strategic alliances. The results provide decision-makers with insights into how they can better assess the long-term value of transport infrastructure investments, as well as offering guidance on how policy frameworks can be adapted to accommodate dynamic territorial development needs.

Furthermore, international case studies offer valuable lessons in transport infrastructure planning, particularly in European contexts, where integrated transport strategies emphasise multimodal connectivity, public transport efficiency, and environmental sustainability. Context-specific approaches are crucial in balancing infrastructure expansion with sustainable urban development practices. Consequetly, this monograph aims to contribute to academic discourse and practical policymaking by systematically analysing decision-making processes in transport infrastructure planning.

Despite its methodological contributions, this research does not claim to offer a definitive solution to the challenge of infrastructure planning. Instead, it presents a flexible and adaptive decision-support mechanism that enhances rational decision-making, improves transparency in complex planning processes, and structures the negotiation of competing interests. Through a critical examination of theoretical frameworks, empirical case studies, and emerging trends, this monograph provides a thorough overview of transport infrastructure planning that is efficient, equitable, and environmentally sustainable.

The subsequent chapters will systematically unfold the research journey. The first section explores the theoretical foundations of decision-support methodologies, followed by an examination of stakeholder participation in territorial governance. The methodological framework is then presented in detail, outlining the temporal multi-criteria decision analysis approach. A comprehensive case study of the Gorizia Interporto illustrates the application of this methodology. The final section evaluates the practical implications and overall effectiveness of the proposed framework. Through this structured analysis, the monograph aims to advance the understanding of transport infrastructure planning, providing actionable insights for policymakers, planners, and researchers engaged in territorial development.

Chapter 2
Strategic Choices in Territorial Development

Abstract In this section, which is dedicated to decision-making in a territorial context, we provide a concise overview of decision support methodologies from the literature. We also explore how the preferences and judgements of multiple individuals can be aggregated into a collective decision. Our focus is on analysising participation in territorial governance within a group decision-making context, using Social Choice Theory as a framework.

Strategic planning and territorial planning involve complex decision-making problems, as it is necessary to simultaneously consider economic, social, and environmental aspects. The objective is to make sustainable choices and better comprehend the long-term consequences of environmental and energy policies, including through a careful analysis of the interactions between physical, biological, and human systems.

The decision-making process becomes even more complex in the domain of transport infrastructure, as it is necessary to consider not only the sustainability of the infrastructures themselves but also the efficacy of the transportation. It is, therefore, essential to balance multiple demands and, consequently, multiple quantitative and qualitative criteria and performance metrics for all the issues addressed [1, 2].

Specifically, transport infrastructures function as a catalyst for urban development, providing accessibility and mobility—essential conditions for efficient and effective ecosystems that enhance local quality of life and foster social innovation aligned with the needs of the community. These needs extend beyond economic growth and the increasing awareness of the imperative for sustainable development. They also encompass social equity and the responsible utilization of natural resources associated with urbanization. All of these considerations must be addressed by governments and stakeholders[1] when planning major transport initiatives.

A comprehensive evaluation of infrastructure performance in terms of sustainability—encompassing economic, social, and environmental dimensions -, renders

[1] The English term refers to the stakeholders in a decision-making process, indicating all those who have a specific interest (*hold*) in an issue, even if they do not have formal decision-making power or explicit legal competence [3].

E. M. Bertolini, *Temporal Value in MCDM*, Lecture Notes in Intelligent Transportation and Infrastructure, https://doi.org/10.1007/978-3-031-95485-6_2

the decision-making process a challenging and complex task. Achieving such projects necessitates balancing multiple quantitative[2] and qualitative[3] performance metrics.

Consequently, transport infrastructure planning issues require an evaluation model that can incorporate all economic, social, and environmental aspects, involving stakeholders and documenting their demands as part of the process itself. Planning, a key component of transportation decision-making processes, is therefore configured as a balancing process of community visions that must be translated into a transport system.

A standard planning process consists of several phases: the first is conceptualization, which identifies problems and various potential solutions; subsequently, project feasibility studies are conducted, through which both the positive and negative aspects of the proposed solutions are evaluated [6].

Strategic Environmental Assessment (SEA)[4] and Environmental Impact Assessment (EIA)[5] are important legal instruments of the European Union's environmental policy that facilitate the identification, prediction, prevention, mitigation, and/or compensation of negative and positive effects. Within the environmental impact

---

[2] Quantitative performance metrics are applicable to facts that have an objective reality, to external data of ongoing processes that can be identified and measured. The relationships between their variables can be determined. The analysis based on them is called quantitative analysis, and it can be statistical, implicative, and factorial, as it classifies characteristics of reality and constructs statistical models to describe the observations made. The results can be directly compared [4].

[3] Qualitative performance metrics are based on an interpretative paradigm, as the facts are strictly dependent on the context and the variables that influence them are interconnected and difficult to measure, and concern the analysis of data that cannot be quantified. The analysis based on them is called qualitative analysis, and it can be descriptive of one or more processes in a specific context, and mostly consists of interviews and discussions. The results are not generalizable and difficult to compare [5].

[4] SEA is the assessment of the effects of certain plans and programs on the natural environment and was introduced in the European Community by Directive 2001/42/EC, the SEA Directive, which came into force on July 21, 2001, and represents an important contribution to the implementation of Community strategies for sustainable development by making the integration of the environmental dimension in strategic decision-making processes operational.

At the national level in Italy, Directive 2001/42/EC was transposed into Part Two of Legislative Decree No. 152 of April 3, 2006, which entered into force on July 31, 2007, and was amended and supplemented by Legislative Decree No. 4 of January 16, 2008, which entered into force on February 13, 2008, and by Legislative Decree No. 128 of June 29, 2010, published in the Official Gazette No. 186 of August 11, 2010.

[5] The EIA aims to ensure that human activity is compatible with the conditions for sustainable development, and therefore in respect of the regenerative capacity of ecosystems and resources, the safeguarding of biodiversity, and an equitable distribution of the benefits of economic activity. In Europe, this procedure was introduced by the Community Directive 85/337/EEC (Council Directive of June 27, 1985, on the assessment of the effects of certain public and private projects on the environment) as a fundamental instrument of environmental policy.

At the national level in Italy, it was transposed into Law No. 349 of July 8, 1986, and subsequent amendments, a law that established the Ministry of the Environment and the rules on environmental damage. With the Decree of the President of the Council of Ministers of December 27, 1988, and subsequent amendments, the Technical Standards for the preparation of Environmental Impact Studies and the formulation of the compatibility judgment were published. The 1985 EIA Directive has been amended five times, in 1997, 2003, 2009, 2011, and 2014.

assessment, variants of the planned activities and investment projects are considered in order to select the most favorable option, including from an environmental perspective. In the context of environmental impact assessment, it is necessary to analyze reasonable alternative design options and select the option where the environmental impact is minimal. The legal requirements do not specify the methods for choosing the optimal option. The ability to make accurate assessments of the environmental impact of a project largely depends on the quality of the information utilized, as well as the application of appropriate selection methods to choose the most favorable option.

The successful implementation of an infrastructure project necessitates clear objectives and multiple criteria for evaluating its feasibility. Multicriteria decision analysis, for instance, is effectively utilized for this purpose and is widely employed in decision-making processes for infrastructure projects [7, 8]. Numerous studies have demonstrated that project decision-makers traditionally consider economic feasibility as the primary decision variable in determining project implementation [8, 9]. This preference results in the neglect of key impacts on a city's level of accessibility, the quality of life of its residents, the maximization of regional economic development, and the reduction of social inequalities [10]. Financial performance receives paramount attention, whereas equal consideration should be given to social and environmental performance, as the construction of transport infrastructures significantly affects the quality of life of residents and the surrounding natural environment.

The practice of prioritizing the economic-financial aspect has led to the predominant use of cost–benefit analysis (CBA) in the feasibility assessment of transport projects, as it facilitates decision-making through direct and clear economic comparisons between alternatives. However, the inherent limitation of cost–benefit analysis is its translation of all non-monetary impacts, such as environmental impact and resident satisfaction, into monetary terms, which can result in controversial decisions [11].

Given the limitations of CBA analysis, there is a need to employ alternative methods, particularly multi-criteria analysis (MCA) methods, which allow for the consideration of both non-monetary and monetary impacts. These approaches are indeed advantageous for project evaluation compared to conventional or single-criterion models.

To address the challenges associated with the integration of sustainability in transport planning, MCA approaches have been extensively utilized and have been recognized as the most appropriate tool for addressing integrated assessment problems [12–14].

2.1 Multicriteria

For many years, a single-criterion approach was utilized to address decision-making problems, operating under the assumption that it was feasible to consolidate multidimensional aspects into a single (economic) objective and reduce them to a singular scale of measurement (monetary). The application of a single-criterion approach, such as focusing exclusively on economic objectives and monetizing all impacts, has been demonstrated to be reductive in addressing complex decision-making problems involving multiple, often conflicting criteria.

Over the past five decades, there has been a shift towards an alternative approach that is more suitable for the evaluation of alternatives characterized by multiple, conflicting criteria. These approaches are classified in the literature as Multi-Criteria Analysis (MCA).

MCA encompasses a family of techniques designed to support the decision-making process in multidimensional problem domains,[6] where multiple alternatives must be evaluated with respect to different evaluation criteria, through various dimensions of those domains. These models have demonstrated a proven track record of providing robust and effective support to decision-makers, forming the foundation for the development of a series of diverse decision support models.

Multi-criteria analysis (MCA) encompasses a wide range of distinct approaches, in contrast to cost–benefit analysis, which is a more established method within a "standard" evaluation framework [16]. All MCA approaches explicitly delineate the options and their contribution to the different criteria, and they all necessitate the exercise of judgment. They differ in the manner in which they combine the data, and the formal MCA techniques typically provide an explicit system of relative weighting for the different criteria adopted.

These MCA approaches can be employed to identify a single preferred option, rank the options, select a limited number of options for further detailed evaluation, or simply to distinguish acceptable from unacceptable possibilities.

There has been a significant proliferation of MCA approaches and related methodological frameworks over the past two decades, as evidenced by an expanding body of literature (see the bibliography, particularly the more recent works such as Broniewicz and Ogrodnik, Dell'Ovo et al., and Huang et al. [17–19].

This underscores the flexibility and adaptability of MCA as a framework for addressing complex, multi-faceted decision-making problems, in contrast to more rigid, single-criterion approaches.

It is essential to note that for approaches to be classified as MCA, they must exhibit certain fundamental characteristics, including a finite set of alternatives (solutions or actions), a minimum of two criteria, and at least one decision-maker.

Considering these characteristics, MCA is an activity that facilitates decision-making primarily in terms of choice, ranking, or ordering of actions [20].

The proliferation of MCA techniques can be attributed to several factors: the diversity of decision-making contexts to which they are applicable; the quantity and nature of available data, which is often heterogeneous and challenging to aggregate. Additionally, there is an increasing recognition among public administrations of the multiple advantages offered by this approach.

---

[6] Integral parts of MCA are Multiple-criteria decision-making (MCDM) and Multi-criteria decision analysis (MCDA) methodologies. These are all subsystems of operational research aimed at explicitly evaluating various conflicting criteria within decision-making processes [15].

These advantages are evident in the capacity of MCA to render the decision-making process explicit and transparent. This transparency is achieved through several means: the selection of objectives and criteria is subject to analysis and modification if deemed inadequate; scores and weights, when utilized, are explicit and developed according to established methodologies; and performance measurement can be delegated to experts, thus not necessitating its execution by the decision-making body itself. The implementation of such approaches can enhance communication within the decision-making body and, in some instances, subsequently between that body and the broader community, while also providing a clear audit trail.

The criteria employed in this context for the selection of techniques are: internal consistency and logical soundness of the model, frequency of utilization, transparency of the decision-making process, and capacity to provide an audit trail.

The selection among different alternatives in multi-criteria analysis involves establishing preferences among various options. These preferences are articulated with reference to an explicit set of objectives identified by the decision-making body, for which measurable criteria have been established to evaluate the extent to which each option achieves those objectives. These objectives can be multiple and/or conflicting and fundamentally represent the specific values of the various stakeholders involved. The articulation of these values in terms of objectives can be beneficial for numerous reasons: for instance, it facilitates the identification, elaboration, and prioritization of alternatives that contribute most significantly to the realization of those values [21].

The definition of objectives and criteria is therefore a key characteristic of MCA, as is the estimation of the relative weights of those criteria. A standard feature of multi-criteria analysis is the creation of a performance matrix in which each row describes an option and each column describes the performance of the options with respect to each criterion (or vice versa). In a basic form of MCA, this performance matrix can be the final product of the analysis, while in more analytically sophisticated MCA techniques, the information in the basic matrix is usually converted into coherent numerical values [22] (Fig. 2.1).

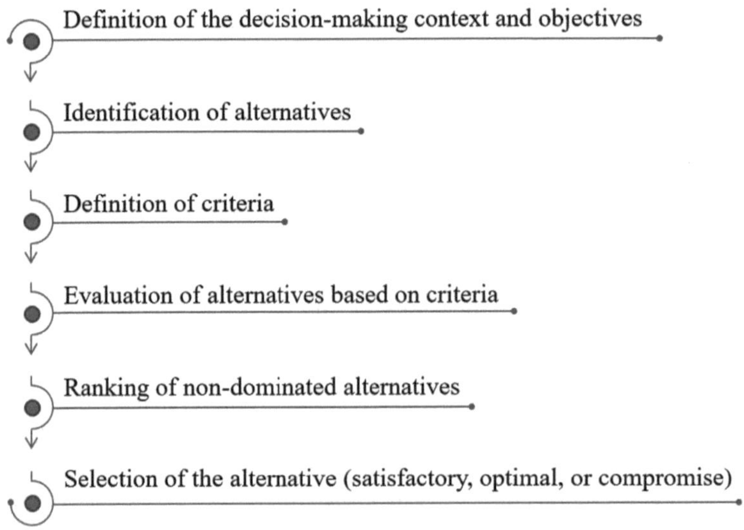

Fig. 2.1 Summary of the key steps in an MCA evaluation. *Source* Author's elaboration

References

1. Boz MA, El-Adaway IH (2014) Managing sustainability assessment of civil infrastructure projects using work, nature, and flow. J Manag Eng 30:4014019
2. Yao H, Shen L, Tan Y, Hao J (2011) Simulating the impacts of policy scenarios on the sustainability performance of infrastructure projects. Autom Constr 20:1060–1069
3. Cartenì A (2016) Processi decisionali e Pianificazione dei trasporti. Lulu.com
4. Campbell DT, Russo MJ (2001) Qualitative research methods in program evaluation. Social Measurement. Thousand Oaks, London, New Delhi, pp 259–286
5. Cardano M (2011) La ricerca qualitativa. Il mulino
6. Gardiner PD (2005) Project management: a strategic planning approach. Macmillan International Higher Education
7. Balali V, Zahraie B, Roozbahani A (2014) A comparison of AHP and PROMETHEE family decision making methods for selection of building structural system. Am J Civ Eng Archit 2:149–159
8. Yun S, Caldas CH (2009) Analysing decision variables that influence preliminary feasibility studies using data mining techniques. Constr Manag Econ 27:73–87
9. Shen L, Tam VWY, Tam L, Ji Y (2010) Project feasibility study: the key to successful implementation of sustainable and socially responsible construction management practice. J Clean Prod 18:254–259
10. Wei H, Asce SM, Liu M et al (2013) Prioritizing sustainable transport projects through multi-criteria group decision making: case study of Tianjin Binhai New Area, China, pp 1–9. https://doi.org/10.1061/(ASCE)ME.1943-5479.0000449
11. Boardman AE, Greenberg DH, Vining AR, Weimer DL (2017) Cost-benefit analysis: concepts and practice. Cambridge University Press
12. Campos VBG, Ramos RAR, de Miranda e Silva Correia D (2009) Multi-criteria analysis procedure for sustainable mobility evaluation in urban areas. J Adv Transp 43:371–390
13. Hickman R, Saxena S, Banister D, Ashiru O (2012) Examining transport futures with scenario analysis and MCA. Transp Res Part A Policy Pract 46:560–575. https://doi.org/10.1016/J.TRA.2011.11.006

14. Šaparauskas J, Turskis Z (2006) Evaluation of construction sustainability by multiple criteria methods. Technol Econ Dev Econ 12:321–326
15. Masotto N (2018) La valutazione ambientale nelle trasformazioni territoriali in ambito alpino: applicazione del Metodo AHP
16. Falvo A, Marabucci A (2008) L'analisi costi benefici applicata alle infrastrutture di trasporto. Department of Economics-University Roma Tre
17. Broniewicz E, Ogrodnik K (2020) Multi-criteria analysis of transport infrastructure projects. Transp Res Part D Transp Environ 83:102351
18. Dell'Ovo M, Dell'Anna F, Simonelli R, Sdino L (2021) Enhancing the cultural heritage through adaptive reuse. A multicriteria approach to evaluate the Castello Visconteo in Cusago (Italy). Sustainability 13 (in press)
19. Huang H, De Smet Y, Macharis C, Doan NAV (2021) Collaborative decision-making in sustainable mobility: identifying possible consensuses in the multi-actor multi-criteria analysis based on inverse mixed-integer linear optimization. Int J Sustain Dev World Ecol 28:64–74
20. Figueira J, Greco S, Ehrott M (2005) Multiple criteria decision analysis: state of the art surveys
21. Keeney RL (1992) Value-focused thinking: a path to creative decisionmaking. Harvard University, Cambridge, MA
22. Baltussen R, Niessen L (2006) Priority setting of health interventions: the need for multi-criteria decision analysis. Cost Eff Resour Alloc 4:1–9

Chapter 3
Review of Decision Support Methodologies in Literature

Abstract This chapter provides an overview of decision support methodologies, with a focus on Multi-Criteria Analysis (MCA) and its application in infrastructure planning. It examines the structure, strengths, and limitations of key MCA methods and assesses their suitability for complex decision-making in the transport and logistics sectors. It also explores the evaluation techniques employed to support strategic choices, emphasising their role in reconciling technical, economic, and policy considerations. Finally, it reviews of contemporary MCA applications in transportation planning, drawing insights from global scientific literature to identify trends, challenges, and best practices in decision support frameworks.

3.1 Basic Concepts and Key Definitions

To better understand the family of methods that fall within the field of multi-criteria analysis, it is necessary to establish the following definitions.

Objective

First, it is important to define the objective. An objective is defined as the indication of the preferred direction of movement such that once the objectives are set, they can be maximized or minimized to achieve the best direction [1]. There are two directions for improvement: one can be interpreted as "more is better" and implies a maximization process; the other can be interpreted as "less is better" and implies a minimization process. Minimizing the inflation rate, CO_2 emissions, or maximizing economic growth are typical examples of objectives within a public policy context.

Target

It is necessary to introduce the concept of target or aspirational level. This represents an acceptable level of achievement for any of the objectives considered by the decision maker. In other words, it indicates the position that the decision-maker considers satisfactory and sufficient.

E. M. Bertolini, *Temporal Value in MCDM*, Lecture Notes in Intelligent Transportation and Infrastructure, https://doi.org/10.1007/978-3-031-95485-6_3

Criterion

Similarly, a judgment criterion is defined as the element to be defined to pursue an objective following a predetermined direction. Is a parameter or evaluation dimension used to compare, measure, and assess the available alternatives in a decision-making process; therefore, it is indispensable to define both the objectives and the criteria through which to pursue them. Each criterion can be quantitative or qualitative, and contributes to determining the effectiveness of each alternative in relation to the overall objectives.

Weight

It indicates the relative importance of criteria in the evaluation process and establishes a hierarchy of importance among evaluation criteria. The weight definition phase, also known as weighting, prevents all criteria from having equal relevance in the assessment. The weight can be expressed as a percentage value, such that the sum of all weights equals 100%. Weighting provides a standardized method for comparing diverse criteria and enables more nuanced and realistic evaluation outcomes. The assignment process can be determined by policy decision makers, established by subject matter experts, or involve the use of structured methodologies for weight determination.

The application is used in the aggregation of multiple criteria scores, helps to create balanced and representative final assessments, and facilitates the incorporation of stakeholder priorities. This weighting system is crucial in multi-criteria analysis, as it allows for a more realistic and nuanced evaluation that reflects the relative importance of different criteria in the decision-making process.

Attribute

The concept of attribute must be defined. It is an expression of decision-maker's values for each criterion. These can be measured independently of the decision-maker's desires (condition of objectivity) and can, in many cases, be expressed as a mathematical function of the underlying decision variables. For example, the inflation rate, economic growth rate, or unemployment rate can be viewed as attributes of economic policy design, whereas pollutant emissions are an example of an attribute in the context of environmental policy.

Attributes have some key characteristics: represent measurable qualities or characteristics, can be objectively quantified, serve as indicators or metrics for evaluating policy or project outcomes.

Goal

The concept of a goal is obtained by combining an attribute with a target. At this point, it is necessary to establish what differentiates a goal from a constraint in a strict sense, given that from a mathematical perspective, both are described by inequalities. In the case of constraints, such inequalities must necessarily be respected; otherwise the alternative must be discarded, while in the case of goals, they represent aspirations of the decision-maker that may or may not be realized.

3.1.1 Formal Mathematical Representation of MCA Concepts

Therefore, MCA refers to the general framework or paradigm for analyzing decision problems involving different attributes, objectives, and goals.

Key distinctions are conducted:

1. Objectives indicate direction (maximize/minimize)
2. Targets represent acceptable achievement levels
3. Goals combine attributes with targets
4. Criteria encompass attributes, objectives, and goals
5. Constraints are mandatory while goals are aspirational.

This structured approach provides a comprehensive framework for analyzing complex decision-making problems with multiple criteria and objectives.

In mathematical terms, an attribute can be represented as a function of the vector of decision variables, i.e.:

$$f(\overline{x}) \tag{3.1}$$

where $\overline{x}$ is the vector of the decision variables.

An objective is represented by the maximization or minimization of the aforementioned function, that is

$$Max\, f(\overline{x})\; o\, Min\, f(\overline{x}) \tag{3.2}$$

where $f(\overline{x})$ is the function of the vector of decision variables.

Finally, an objective corresponding to the achievement of a target value $\hat{f}$ is represented as

$$f(\overline{x}) + n - p = \hat{f} \tag{3.3}$$

where n is the negative deviation variable, i.e., the quantification of the failure to achieve the objective, p is the positive deviation variable, i.e., the quantification of the exceeding of the objective, $\hat{f}$ is the target value to be achieved.

Therefore, each alternative is assigned a score that will be presented to the decision-maker, and the measured value will identify the value of the alternative [2] (Fig. 3.1).

Formal mathematical framing provides a rigorous and standardized way to represent key MCA concepts, enabling their systematic application in decision-making processes.

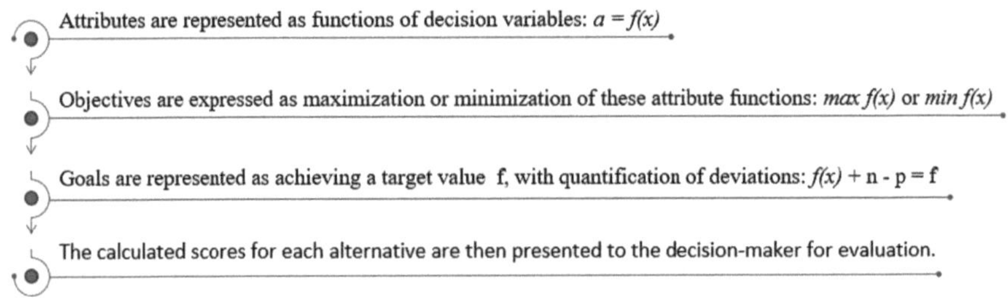

Fig. 3.1 Key points of a MCA

3.1.2 *The Pareto Optimum*

The concept of Pareto optimality has been successfully transferred from economics to Multi-Criteria Analysis (MCA) by simply substituting the entity "person" (economics) with the entity "criterion" (MCA). It has recently been proposed as an interpretation of collective choices, serving as a tool for seeking a unanimous agreement on public expenditure among a group of subjects with their own preferences.

According to the Pareto criterion, in a community composed of H individuals with individual utility functions $U^h(x^h)$ con $h = 1, 2, \ldots, H$, which are orderable but not comparable, defined in the individual basket of goods $x^h = \{x_1^h, \ldots, x_N^h\}$, the state of the economy is defined as the set of individual baskets:

$$x = \{x^1, \ldots, x^h, \ldots, x^H\} \tag{3.4}$$

where x^h represents the basket of goods for each individual.

A state of economy x* is Pareto efficient if there exists no other state of economy x^+ such that:

$$U^h(x^+) \geq U^h(x^*) \tag{3.5}$$

for all individuals $h = 1, \ldots,$ H, where U^h represents an individual's utility function and x^* and x^+ represent two different states of the economy.

Because the Pareto criterion is used to identify collective choices based on individual utility functions, Pareto optimality is achieved when resource allocation reaches a point where it is impossible to improve the system further, that is, one cannot improve one condition without deteriorating another.

In MCA, Pareto-efficient or optimal solutions are those that are acceptable across all criteria under examination and are strictly better for at least one criterion. Naturally, all MCA techniques aim to obtain solutions that are at least Pareto efficient, as defined above.

The concept of Pareto optimality can lead to the notion of trade-offs between two criteria and measures the level of one criterion that must be sacrificed to achieve a unit improvement in another criterion. Furthermore, these values are crucial for the design and evaluation of any economic and environmental policy [3].

3.2 MCA Methods

MCA models are divided into multi-objective and multi-attribute evaluation methods [4]:

- Multiple Objective Decision Making (MODM). MODM is predominantly utilized for design or planning purposes, where the number of alternatives is infinite. The objective of MODM is to design the optimal alternative by considering various interactions within given constraints. The optimal alternative is achieved by attaining satisfactory levels across multiple objectives.
- Multiple Attribute Decision Making (MADM). MADM is primarily implemented for selection problems and is invariably associated with a finite number of alternatives and preference rankings. A finite set of proposed alternatives is evaluated with respect to various weighted attributes, resulting in a preference ranking that describes the performance of each alternative in meeting the objective relative to the attributes.

3.2.1 Multi-objective Methods

Multi-objective methods are characterized by a probabilistic approach owing to the infinite number of alternatives, with their primary aim being to identify the alternative that ensures the achievement of acceptable levels across multiple objectives.

While many decisions are driven by a single objective, decision-makers frequently encounter situations requiring the satisfaction of multiple objectives, which may be either complementary or conflicting in nature. In the case of complementary (non-conflicting) objectives, a single alternative can simultaneously satisfy multiple objectives; the preferred alternative will be the one that maximizes the greatest number of possible objectives. When objectives conflict, competition for available resources emerges, as these resources can be allocated to one objective or another but not both simultaneously. The precise manner of competition and the basis for the dominance of one objective over another depend on the nature of the developed decision rule.

In scenarios involving complementary objectives, multi-objective decisions can often be resolved through a hierarchical extension of the multi-criteria evaluation process [5].

Given the definition and concept of the term "objective" in the literature, it is essential to clarify its meaning within this context. An objective represents a statement of what the decision-maker, whether an individual or group, hopes to resolve the problem at hand. Furthermore, for consistency and clarity, the objectives should be expressed in the form of a verb followed by an object. Examples of these objectives include maximizing profits, increasing product sales, limiting environmental degradation, reducing dependency on foreign energy sources, enhancing educational quality, and decreasing vehicle fatalities.

Identifying an appropriate set of objectives for a complex decision is not trivial and requires careful consideration. To facilitate this process, MOD are treated as natural extensions of mathematical programming, in which multiple objective functions are considered simultaneously.

The general formulation of a Multiple Objective Programming problem (MOP, sometimes referred to as a "vector maximum problem") is:

$$\begin{gathered} \max\ [g_j(\mathrm{X}), \ldots, g_p(\mathrm{X})] \\ \text{s.t. } h_i(\mathrm{X}) > 0\ (i = 1, \ldots, m) \\ + \text{ additional constraints (integrity constraints, \ldots).} \end{gathered} \tag{3.6}$$

where g_j (X) (j = 1,..., p) represent the objective functions and are real-valued functions, h_i (X) (i = 1,..., m) represent the constraint functions.

The notation "max" indicates the search for the best compromise for a Decision Maker (DM) based on its preference structure, considering each of the p criteria (it is assumed, without loss of generality, that each criterion is to be maximized).

A special case is the MOP problem formulated as follows:

$$\begin{gathered} \max\ [\mathrm{c}^{\mathrm{i}}\mathrm{X}, \ldots,\ \mathrm{c}^{\mathrm{p}}\mathrm{X}] \\ \text{s.t. } D\mathrm{X} \leq \mathrm{b} \\ \mathrm{X} \leq 0 \end{gathered} \tag{3.7}$$

where c^{i} are n-dimensional coefficient vectors, D is an $m \times n$ constraint coefficient matrix, b is an m-dimensional vector.

The general problem is frequently reformulated in the criteria space using variables z_j which denote the values of criteria g_j;

$$\begin{gathered} \max\ [\mathrm{z}_1, \ldots, \mathrm{z}_\mathrm{p}] \\ \text{s.t.} z_j = g_j(a)\ (j = 1, \ldots, p) \\ a \in A. \end{gathered} \tag{3.8}$$

where g_{j} represent the criteria, z_{j} denote the criteria values, A denotes the set of potential (or feasible) alternatives (a).

Building upon these formulations, a set of conventional concepts has been developed in the literature, with the most prevalent being "non-dominated" (or "weakly non-dominated") solutions, expressed in the criteria space, and their corresponding "efficient" (or "weakly efficient") solutions in the decision variable space [6].

3.2.1.1 TOPSIS

TOPSIS (Technique for Order of Preference by Similarity to Ideal Solution) is a straightforward and yet effective evaluation method based on in the concept of compromise. It identifies the optimal alternative among multiple feasible options by calculating distances from both positive ideal and negative ideal (anti-ideal) solutions [7]. This approach is widely implemented in solving Multi-Criteria Decision Making (MCDM) problems. Initially proposed as an applicable MCDM approach by [4] and subsequently expanded by [8], TOPSIS offers several advantages: intuitive and transparent decision-making logic, computational efficiency, the capability to measure each alternative's appropriateness relative to ideal solutions, and visual representation of results [9].

A single action or alternative (A_i) can be represented as:

$$A_i = \left(x_{i1}, \ldots, x_{ij}, \ldots, x_{in}\right),\ i = 1, \ldots, n \quad (3.9)$$

where x_{ij} represents the performance of the ith alternative with respect to the jth attribute.

The positive ideal solution is defined as the aggregation of all the optimal achievable values for each attribute, whereas the negative ideal solution comprises all the worst achievable values for each attribute.

The TOPSIS method consists of the following steps:

1. Create a decision matrix $X = (x_{ij})_{mxn}$ comprising m alternatives and n criteria

$$X = \begin{bmatrix} x_{11} & x_{12} & \cdots & x_{1n} \\ x_{21} & x_{22} & \cdots & x_{2n} \\ \vdots & \vdots & x_{st} & \vdots \\ x_{m1} & x_{m2} & \cdots & x_{mn} \end{bmatrix} \quad (3.10)$$

2. Normalize and weight the decision matrix, where a_{ij} represents the normalized value of the ith alternative with respect to the jth attribute

$$a_{ij} = \frac{x_{ij}}{\sqrt{\sum_{i=1}^{m} x_{ij}^2}},\ i = 1, 2, \ldots, m; j = 1, 2, \ldots n \quad (3.11)$$

3. Weight the normalized decision matrix by defining the weighting coefficients for each attribute. Note that the method for determining weights is flexible—they can be assigned either arbitrarily or based on decision-makers' judgments

$$r_{ij} = a_{ij} \cdot w_j,\ i = 1, 2, \ldots, m; j = 1, 2, \ldots n \quad (3.12)$$

4. Determine the ideal solution matrices (both positive and negative)

$$A^{+} = \left\{r_1^{+}, r_2^{+}, \cdots, r_n^{+}\right\} = \left\{\left(\max r_{ij} \lfloor j\epsilon I'\right), \left(\min r_{ij} \lfloor j\epsilon I''\right), \right\} \tag{3.13}$$

$$A^{-} = \left\{r_1^{-}, r_2^{-}, \cdots, r_n^{-}\right\} = \left\{\left(\min r_{ij} \lfloor j\epsilon I'\right), \left(\max r_{ij} \lfloor j\epsilon I''\right)\right\} \tag{3.14}$$

5. Calculate the distance between each alternative's value and both the positive and negative ideal solution matrices:
   For the distance D_i^+ between alternative A_i and the positive ideal solution.
   For the distance D_i^- between alternative A_i and the negative ideal solution.

$$D_i^{+} = \sqrt{\sum_{j=1}^{n} \left(r_{ij} - r_j^{+}\right)^2}, i = 1, 2, \ldots, m; j = 1, 2, \ldots n \tag{3.15}$$

$$D_i^{-} = \sqrt{\sum_{j=1}^{n} \left(r_{ij} - r_j^{-}\right)^2}, i = 1, 2, \ldots, m; j = 1, 2, \ldots n \tag{3.16}$$

6. Calculate the relative closeness coefficient (C_i^+), for the ith alternative, which represents the distance between the case under examination and the two artificial solutions.

$$C_i^{+} = \frac{D_i^{-}}{D_i^{+} + D_i^{-}}, i = 1, 2, \ldots, m; \tag{3.17}$$

7. Rank preferences in descending order. The optimal alternative corresponds to the highest value of di C_i^+ as if $C_i^+ = 1$ then $A_i = A^+$.

For a proper implementation of this methodology, the values must be numerical, monotonically increasing or decreasing, and characterized by commensurable units [10]. In 1965, Zadeh proposed the use of fuzzy sets to numerically represent linguistic terms and manage uncertainties inherent in human judgment [11]. In engineering applications, this is typically achieved using the Gumus scale, which quantifies the relative importance of the elements through numerical triplets. In these triplets, the central number represents the value on Saaty's semantic scale, whereas the other two numbers represent the minimum and maximum possible values [12]. The TOPSIS method has been deemed advantageous owing to both its relatively straightforward computational algorithm and capability to analyze quantitative data [9] (Fig. 3.2).

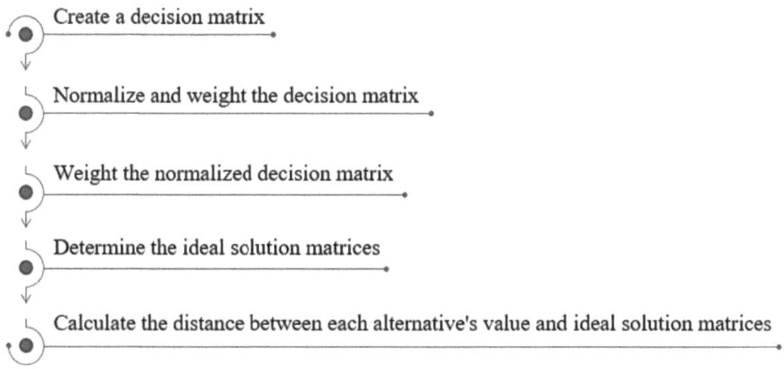

Fig. 3.2 The TOPSIS method step

3.2.2 Multi-attribute Methods

In contrast to multi-objective approaches, multi-attribute methods analyze a finite number of alternatives by employing a deterministic approach, owing to their predetermined set of options. The primary objective of these methods is to identify alternatives associated with a satisfactory level of attribute fulfillment.

Most multi-attribute methods construct an aggregation of evaluations associated with each attribute, enabling the derivation of a unified preference structure across the entire set of compared elements. This allows the ranking of predefined alternatives based on decision-maker preferences. The multi-attribute approach allows for structuring the decision-making process systematically, considering the complexity of the real-world choices.

In decision theory, multi attributes models are mainly divided into two major families:

- Objective Function Models (also known as utility models). These models are based on the construction of a mathematical function that aggregates different criteria into a single numerical value. The fundamental idea is that it is possible to express the overall value of an alternative using a mathematical function that combines the scores of individual attributes. Decision-maker preferences are modeled using weights and utility functions. The most well-known method is MAUT (Multi-Attribute Utility Theory), but there are also simplified approaches, such as the weighted sum method. The key assumption is that perfect compensation exists between criteria.
- Outranking Models. These models, developed mainly by the French school, are based on a different approach that does not assume complete compensation between the criteria. The key concept is "outranking", when one alternative outranks another if it is at least as good in most criteria and not too much worse in others. These methods accept the incomparability between alternatives and use preference and indifference thresholds to model uncertainty in evaluations. The most well-known methods are ELECTRE and AHP models. They are particularly

suitable when it is not possible or desirable to aggregate all the criteria into a single numerical value.

3.2.2.1 Objective Function Models

Linear Additive Models

Linear additive models are constructed by combining the values of an option across multiple criteria into an overall value. This is achieved by multiplying the value score for each criterion by its corresponding weight and subsequently summing all weighted scores.

These models are applicable in scenarios where preference criteria demonstrate mutual independence, and uncertainty is not formally incorporated into the Multi-Criteria Analysis (MCA) framework. Most MCA approaches employ this additive model, which serves as the foundation for different models.

Models of this type have a proven track record in providing robust and effective decision support across a diverse range of problems and circumstances, forming the basis for the development of various decision support frameworks.

Simple Additive Weighting Method

The *Simple Additive Weighting* (SAW) method can be considered the most intuitive and straightforward approach to addressing MCA decision problems, as its linear additive function can effectively represent decision-makers' preferences. However, this holds true only under the assumption of preference independence [1] or preference separability [13].

Churchman and Ackoff first employed the SAW method to address a portfolio selection problem [14]. Owing to its simplicity, the SAW method is arguably the most well-known and widely utilized method for multiple criteria decision-making. It identifies the optimal alternative using the following equation:

$$A^* = \{u_i(x) | max_i u_i\ (x) | i = 1, 2, \ldots, n\} \tag{3.18}$$

where A* represents the preferred alternative, $i = 1, 2, \ldots, n$ denotes the n alternatives, $u_i\ (x)$ represents the utility of the ith alternative.

Furthermore, this can be expressed as:

$$u_i(x) = \sum_{j=1}^{n} w_j r_{ij}(x) \tag{3.19}$$

where $u_i(x)$ represents the utility of the ith alternative, w_j denotes the weight of the jth criterion, $r_{ij}(x)$ represents the normalized preference rating of the ith alternative with respect to the jth criterion.

In this context, all the criteria are presumed to be independent.

The normalized evaluations ($r_{ij}(x)$) of the ith alternative with respect to the jth criterion can be arranged according to two models.

Model 1

- For benefit criteria (larger is better):

$$r_{ij}(x) = \frac{x_{ij}}{x_j^*} \tag{3.20}$$

where x_{ij} represents the value of the ith alternative for the jth criterion, x_j^* represents the optimal scenario, such that $0 \leq r_{ij}(x) \leq 1$;
- For the cost criteria (smaller, better),

$$r_{ij}(x) = \frac{\frac{1}{x_{ij}}}{\frac{1}{x_j^*}} = \frac{\max x_{ij}}{x_{ij}} \tag{3.21}$$

where x_{ij} represents the value of the ith alternative for the jth criterion, $x_j^* = max\ x_{ij}$ represents the optimal scenario.

Model 2

- For benefit criteria (larger is better):

$$r_{ij}(x) = \frac{(x_{ij} - x_j^-)}{x^* - x_j^-} \tag{3.22}$$

where x_{ij} represents the value of the ith alternative for the jth criterion, $x_j^* = max\ x_{ij}$ o x_j^* r represents the value of the best-performing alternative for the jth criterion, x^-_j represents the value of the least-performing alternative for the jth criterion.
- For cost criteria (smaller is better):

$$r_{ij}(x) = \frac{x_j^- - x_{ij}}{x_j^- - x_j^*} \tag{3.23}$$

where x_{ij} represents the value of the ith alternative for the jth criterion, x^-_j represents the value of the least-performing alternative for the jth criterion, $x^* = max\ x_{ij}$ represents the optimal scenario.

Regardless of the model employed, the synthesized performance is:

$$p_i = \sum_{j=1}^{m} w_j r_{ij} \tag{3.24}$$

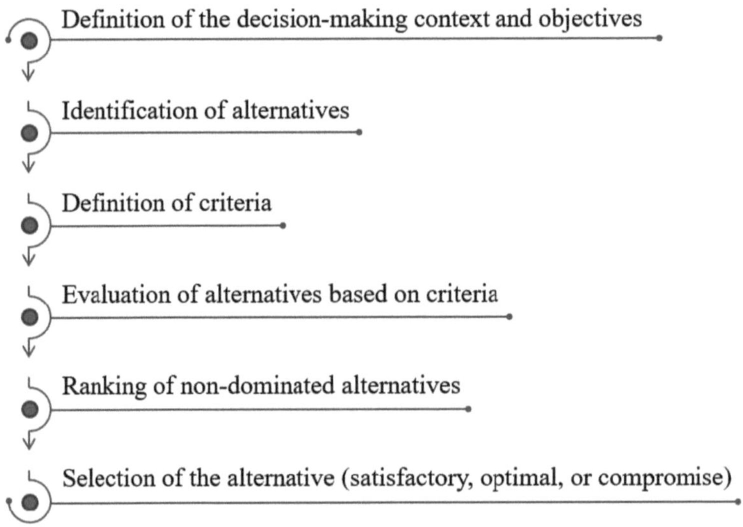

Fig. 3.3 Phases of the decision-making process in the SAW framework. *Source* Author's elaboration

where p_i is the synthetic performance value of the ith alternative, w_j indicates the weights of the jth criterion, r_{ij} is the normalized evaluation of the ith alternative with respect to the jth criterion.

Furthermore, the criteria are assumed to be mutually independent [15] (Fig. 3.3).

Multi Attribute Value Theory

Multi Attribute Value Theory (MAVT) is a methodological framework that provides prescriptive decision-making recommendations for choosing between alternatives $x = (x_1, \ldots, x_n)$ that have consequences across n attributes [16–18]. MAVT is founded on a set of axioms that characterize rational decision-making. For instance, it postulates that a rational decision-maker has complete preferences, meaning that for any two given alternatives x and y, the decision-maker either finds these alternatives equally preferable or prefers one over the other. Furthermore, preferences are assumed to be transitive, meaning that if the decision-maker prefers alternative x over y and alternative y over z, then x is logically preferred over z.

Preferential mutual independence is a key axiom in MAVT [18]. Specifically, this axiom holds if the decision-maker's preferences for alternatives, which have different consequences on some attributes and similar consequences on other attributes, remain unchanged when the similar consequences of the alternatives are modified. If this axiom holds along with other less restrictive axioms, there exists an additive multi-attribute value function such that alternative x is preferred to y if and only if:

$$x \geq y \iff V(x) = \sum_i v_i(x_i) \geq \sum_i v_i(y_i) = V(y) \tag{3.25}$$

where x and y are two alternatives; V(x) and V(y) are the value functions of alternatives x and y, respectively; $v_i(x_i)$ is the value of the single attribute for alternative x; $v_i(y_i)$ is the value of the single attribute for alternative y.

The existence of the value function has been demonstrated using both topological [19] and an algebraic approach [20]. The value function is unique up to positive affine transformations. Therefore, the preference relationship it induces on alternatives remains unchanged if the values are multiplied by a positive constant $\alpha > 0$ or if a constant β is added to the overall values of all alternatives. Due to this property, the MAVT function expressed above can be written in its conventional form:

$$V(x) = \sum w_i v_i(x_i) \tag{3.26}$$

where V(x) is the value function of alternative x, $v_i(x_i)$ is the value of the single attribute for alternative x, w_i indicates the weight of the ith attribute.

In this case, the scores $v_i(x_i)$ are typically normalized within the range [0, 1], assigning zero to the least preferred alternative and one to the most preferred alternative. Furthermore, w_i denotes the attribute weights, which reflect the decision-maker's preferences and thus modify their choices. These weights are normalized so that their sum equals 1, i.e., $\sum w_i = 1$.

Keeney and Raiffa extend the MAVT framework to accommodate group decision-making contexts where the group's aggregate value depends on the values achieved by individual group members. Specifically, they demonstrate that if the required axioms hold, the group's aggregate value function can be expressed as:

$$V(x) = \sum_k W_k \sum w_{ki} v_{ki}(x_i) \tag{3.27}$$

where V(x) is the value function of alternative x; $v_{ki}(x_i)$ is the value of the single attribute for alternative x; w_{ki} indicates the weight of the ith attribute; W_k denotes the importance weight of the kth decision-maker, and this sum represents the value that alternative x will provide to them [21] (Fig. 3.4).

When implementing the MAVT framework in group decision making support, the representation parameters are first estimated; subsequently, the overall values of alternatives are used to derive decision recommendations. One of the primary advantages of the MAVT framework is its solid and verifiable axiomatic foundation. Moreover, the numerical representation is relatively straightforward, making MAVT models quite transparent, thus facilitating understanding of how decision recommendations depend on the estimated parameters [23].

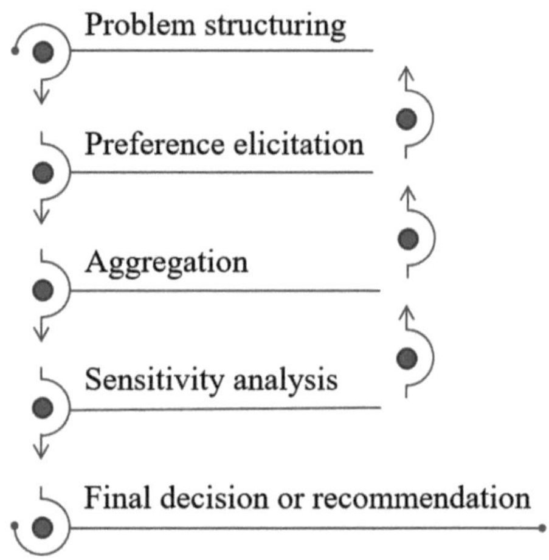

Fig. 3.4 Phases of the decision-making process in the MAVT framework. *Source* Author's elaboration on [22]

3.2.2.2 Outranking Models

Analytic Hierarchy Process

The method developed by Saaty [24] enables the evaluation of intervention strategy priorities by combining tools such as "pairwise comparison", "eigenvector calculation" and consistency judgments. The Analytic Hierarchy Process (AHP) involves decomposing the decision problem into a hierarchy of criteria, sub-criteria, attributes, and alternatives. This hierarchy is combined with a series of weights that reflect the relative importance of components, ultimately allowing the decision-maker to identify the optimal or compromise solution (Fig. 3.5).

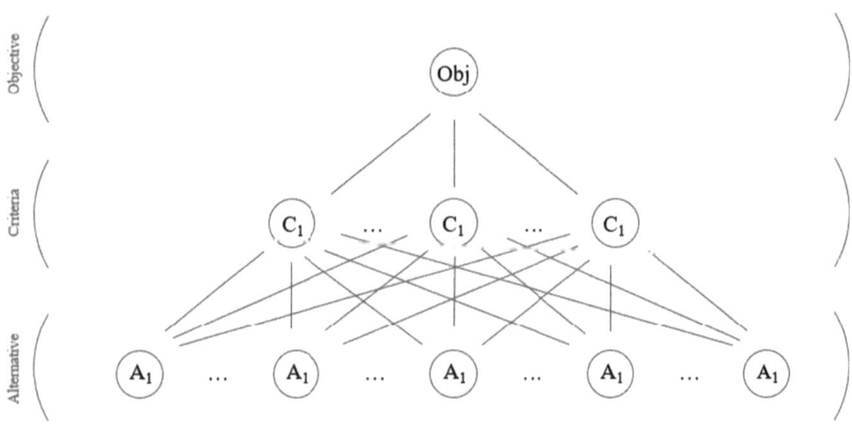

Fig. 3.5 Hierarchy dominance structure. The main objective is positioned at the apex, with supporting criteria (which may be further divided into sub-criteria) positioned below. The alternatives to be evaluated are situated at the base of the hierarchy. *Source* Author's elaboration

The hierarchical structure derived from this decomposition shapes a tree diagram with distinct levels and inter-level connections: the main objective appears at the top of the diagram, branching into sub-criteria, which subsequently lead to additional sub-criteria. This decomposition process continues until reaching the penultimate level of the diagram, where attributes detailing the decision-making process are located, while the alternatives under consideration occupy the final level. The construction of this diagram necessitates the involvement of experts, decision-makers, and potentially the public. Through the succeeding pairwise comparison, a set of weights is identified. It represents the relative importance of criteria, sub-criteria, and attributes. Once these weights are established, a final weight can be assigned to each option, enabling classification.

Upon completion of the hierarchy, the problem can be analyzed with greater effectiveness and efficiency. This is possible because of the property of internal independence among elements at the same level and the property of external dependence between elements of one level relative to those of a superior level. The hierarchy also enables monitoring of the problem-solving process while providing a detailed view of the problem itself [25] (Fig. 3.6).

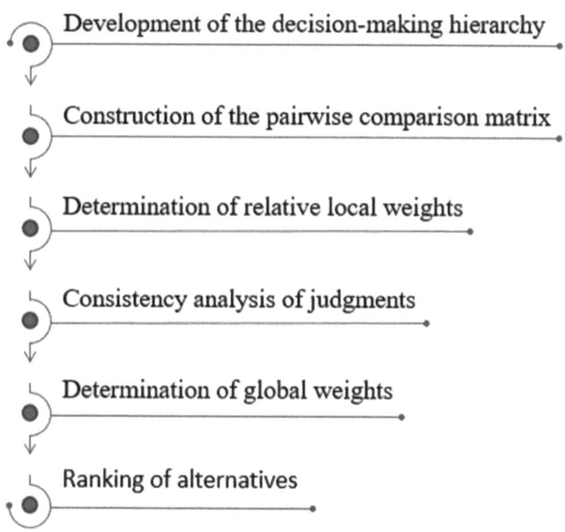

Fig. 3.6 Representative outline of the fundamental phases of the AHP method. *Source* Author's elaboration

3.3 Analysis and Evaluation Methods Used in the Transport and Logistics Sector

The construction and operation of transport infrastructure are fundamental to the effective functioning of economy and society, requiring substantial financial and spatial resources while generating various environmental impacts. Although efficient and operational transport infrastructure is essential and may justify its typically associated negative effects—such as loss of local ecosystems, landscape alterations, air pollutant emissions, noise pollution, and changes in hydrological processes. Thus, remains a critical need to analyse reasonable alternative design options and select those with minimal environmental impact.

The European Union, as part of its environmental policy framework, has adopted legal instruments to identify, predict, and mitigate these negative effects, specifically the Strategic Environmental Assessment (SEA) and Environmental Impact Assessment (EIA). However, legal requirements do not specify methodologies for selecting optimal options. This selection process must be based on reliable information—as the ability to conduct accurate environmental impact assessments largely depends on the quality of information utilized—and appropriate selection methods to identify the most favourable option. Subjective evaluation of options and incorrect analytical methods can lead to suboptimal decisions with negative environmental and societal consequences.

Extensive literature demonstrates the theoretical application of Multi-Criteria Analysis (MCA) methods in resolving transport-related decision problems and selecting environmentally optimal route variants. Analysis reveals that the most widely employed methods for solving multi-criteria decision problems in transportation are modified AHP, TOPSIS, and methods within the European trend, namely PROMETHEE and ELECTRE. Furthermore, MCA methods continue to evolve, and due to the limitations of individual methods, hybrid approaches combining different methods or their algorithms are increasingly prevalent in research.

While MCA is more frequently applied to transport projects rather than policies or programs, several applications of multi-criteria analysis in the transport sector are documented by Tsamboulas and Mikroudis [26], Bouwman and Moll [27] and Yedla and Shrestha [28]. For instance, Portugal's High-Speed Railway Authority (HRSA) conducted an MCA to evaluate railway positioning near Porto's metropolitan centre, ranking alternative locations based on their overall value. In such circumstances, this model was preferred over Cost–Benefit Analysis (CBA) as not all criteria could be readily monetized due to their exogenous and intangible nature [29].

The United Kingdom developed a New Approach to Appraisal (NATA) for evaluating transport projects and proposals, establishing a multi-criteria decision framework based on five criteria: economic, safety, environmental, accessibility, and integration. This framework is organized in an Appraisal Summary Table (AST), which serves as the foundation for evaluating multimodal studies, including road and public transport systems, seaports, and airport strategies [30, 31].

Bristow and Nellthorp reviewed various transport evaluation methods used across European Union (EU) Member States and found strong consensus on handling direct impacts, although less agreement exists regarding the treatment of social impacts [32].

A notable MCA application is the recently completed research project PROSPECTS (Procedures for Recommending Optimal Sustainable Planning of European City Transport Systems) for the European Union, which aimed to identify the "ideal" decision-making process for sustainable transport planning choices in the European context [33]. Subsequently, the EU-funded SPARKLE project (Sustainability Planning for Asian Cities making use of Research, Know-How, and Lessons from Europe) examined the relevance of the PROSPECTS process in Southeast Asia. Analysis of both projects' conclusions highlights key factors influencing evaluation: diverse planning traditions; varying weights in transport objectives; limited use of policy instruments; rapid growth rates (both economic and travel-related); differences in vehicle types; and lack of data for evaluation and modelling [34].

Another practical application involves multi-criteria analysis for establishing a hierarchy of potential levers for territorial logistic enhancement, according to the theoretical dimensions of Auto-regulation, Compensation, Invariance, and Transversality (ACIT), as identified in recent Economic Logistics literature [35].

The AHP method was initially employed as an analytical approach for evaluating aspects of technical innovation in construction, such as the selection of tower cranes [36], or comparing the use of composite materials versus traditional materials in construction, assessing costs and benefits in non-economic terms [37]. A notable application includes the use of AHP for defining values and selected criteria to evaluate concrete structures' conditions through non-destructive testing [38].

In the infrastructure domain, AHP has been applied to highway alignment selection. In this context, the following primary objectives were considered for the highway project under examination: minimization of environmental impact, optimization of alignment functionality (including safety aspects), minimization of construction time (relative to the 60-month deadline), minimization of construction and operational costs, and maximization of economic investment returns [39].

At the territorial level, the AHP method has been implemented in various circumstances, such as the development of a landslide susceptibility index and corresponding mapping of sensitive areas in the Tinau watershed, Western Nepal, in this case, all factors influencing ground instability were classified and subsequently integrated using the weighted linear sum procedure [40]. Bitafataran, in selecting the optimal design scenario for intelligent bridge structures in potential earthquake situations to reduce seismic stress and increase reliability. Five indices were considered: reduction of casualties and vulnerabilities, feasibility of sensor technology localization, performance costs, performance speed, and maintenance. The pairwise comparison method was utilized to determine the final weight of these indices [41].

Broniewicz and Ogrodnik examined the application of multi-criteria methods to select the most environmentally favourable route variant for a highway section in northeastern Poland, comparing their analysis results with the choice made in the environmental impact report. They employed a hybrid approach combining the most commonly applied MCA methods in transport decision problems and used the classical AHP method to estimate criteria weights. In addition to classical AHP, they utilized fuzzy AHP due to the growing significance of fuzzy set theory in multi-criteria methods. Their primary objective was to conduct a comparative analysis of two weight sets and determine the impact of fuzzy scale on factor weights and subsequent decision rankings. Both classical AHP and fuzzy AHP were used for factor weighting, while TOPSIS and PROMETHEE methods were employed for developing final rankings [42].

Evidence suggests that Multi-Criteria Analysis (MCA) methods serve as vital decision support tools in addressing various transport-related challenges, both for selecting transport system development scenarios and analyzing transport project locations. Multi-criteria decision support and its tools constitute a broad field offering numerous opportunities for further research development, as MCA methods continuously evolve, evidenced by the significant and annually increasing number of publications on this topic. While these methods facilitate decision-making in situations characterized by data uncertainty and ambiguity, increasingly complex algorithms may present limitations to their practical application. The implementation of such methods also requires, where possible, precise quantitative data, which in turn demands the expertise of multiple specialists and specialized software.

However, Marsden and Reardon, through their analysis of 100 sampled articles from two major policy journals in transport literature, applying two different frameworks for policy understanding, found that only 13% of papers consider specific aspects of the policy cycle, 60% focus on policy "instruments", and two-thirds of the documents concentrated solely on quantitative analysis without addressing real-world policy examples or policymakers. These authors argue that these findings highlight the persistence of the technical-rational model within transport literature.

The substantial lack of engagement with governance issues and debates suggests the artificial and disproportionate generation of an applied decision-making science that is unlikely to be utilized, due to the disconnect between theoretical frameworks and practical realities in the field [43].

References

1. Keeney RL, Raiffa H, Meyer RF (1993) Decisions with multiple objectives: preferences and value trade-offs. Cambridge University Press
2. Barichard V, Ehrgott M, Gandibleux X, T'Kindt V (2009) Designing public policies an approach based on multi-criteria analysis and computable general equilibrium modeling. Springer
3. Sancho F (2011) Designing public policies. an approach based on multi-criteria analysis and computable general equilibrium modeling. Springer Science & Business Media
4. Hwang C-L, Yoon K, Hwang C-L, Yoon K (1981) Methods for multiple attribute decision making. In: Multiple attribute decision making: methods and applications a state-of-the-art survey, pp 58–191
5. Eastman JR, Jiang H, Toledano J (1998) Multi-criteria and multi-objective decision making for land allocation using GIS. In: Multicriteria analysis for land-use management. Springer, pp 227–251

6. Colson G, De Bruyn C (1989) Models and methods in multiple objectives decision making. Pergamon Press plc
7. Mahdevari S, Shahriar K, Esfahanipour A (2014) Human health and safety risks management in underground coal mines using fuzzy TOPSIS. Sci Total Environ 488–489:85–99. https://doi.org/10.1016/J.SCITOTENV.2014.04.076
8. Chen S-J, Hwang C-L (1992) Fuzzy multiple attribute decision making methods. In: Chen S-J, Hwang C-L (eds) Fuzzy multiple attribute decision making: methods and applications. Springer, Berlin Heidelberg, pp 289–486
9. Roszkowska E (2013) Rank ordering criteria weighting methods–a comparative overview. Optimum Studia Ekonomiczne 14–33
10. Behzadian M, Khanmohammadi Otaghsara S, Yazdani M, Ignatius J (2012) A state-of the-art survey of TOPSIS applications. Expert Syst Appl 39:13051–13069. https://doi.org/10.1016/J.ESWA.2012.05.056
11. Zimmermann HJ (2010) Fuzzy set theory. Wiley Interdiscip Rev Comput Stat 2:317–332. https://doi.org/10.1002/WICS.82
12. Gumus AT (2009) Evaluation of hazardous waste transportation firms by using a two step fuzzy-AHP and TOPSIS methodology. Expert Syst Appl 36:4067–4074. https://doi.org/10.1016/J.ESWA.2008.03.013
13. Gorman WM (1968) Conditions for additive separability. Econometrica 605–609
14. Churchman CW, Ackoff RL (1954) An approximate measure of value. J Oper Res Soc Am 2:172–187
15. Tzeng G-H, Huang J-J (2011) Multi attribute decision making
16. Belton V, Stewart TJ (2002) MCDA in a broader context. In: Multiple criteria decision analysis. Springer, pp 293–329
17. French S (1996) Multi-attribute decision support in the event of a nuclear accident. J Multi-Criteria Decis Anal 5:39–57
18. Keeney RL, Raiffa H, Rajala DW (1979) Decisions with multiple objectives: preferences and value trade-offs. IEEE Trans Syst Man Cybern 9:403. https://doi.org/10.1109/TSMC.1979.4310245
19. Debreu G (1960) Individual choice behavior: a theoretical analysis
20. Krantz D, Luce D, Suppes P, Tversky A (1971) Foundations of measurement, Vol. I: Additive and polynomial representations
21. Meyer RF (1976) Preferences over time. In: Keeney RL, Raiffa H (eds) Decisions with multiple objectives: preferences and value tradeoffs
22. Raskob W, Hugon M, Hiete M et al (2010) Evaluation strategies for nuclear and radiological emergency and post-accident management. Radioprotection 45:S133–S147
23. Kilgour DM, Eden C (2010) Handbook of group decision and negotiation. Springer Science & Business Media
24. Saaty Thomas L (1990) The analytic hierarchy process: planning, priority setting, resource allocation. North-Holland
25. Falcone D, De Felice F, Saaty TL (2009) Il decision marketing ei sistemi decisionali multicriterio. Le metodologie AHP e ANP. HOEPLI EDITORE
26. Tsamboulas D, Mikroudis G (2000) EFECT–evaluation framework of environmental impacts and costs of transport initiatives. Transp Res D Transp Environ 5:283–303
27. Bouwman ME, Moll HC (2002) Environmental analyses of land transportation systems in The Netherlands. Transp Res D Transp Environ 7:331–345
28. Yedla S, Shrestha RM (2003) Multi-criteria approach for the selection of alternative options for environmentally sustainable transport system in Delhi. Transp Res Part A Policy Pract 37:717–729
29. Mateus R, Ferreira JA, Carreira J (2008) Multicriteria decision analysis (MCDA): Central Porto high-speed railway station. Eur J Oper Res 187:1–18
30. Glaister S (1999) Observations on the new approach to the appraisal of road projects. JTEP 33:227–234
31. Price A (1999) The new approach to the appraisal of road projects in England. JTEP 33:221–226

32. Bristow AL, Nellthorp J (2000) Transport project appraisal in the European Union. Transp Policy 51–60
33. Järvi-Nykänen T (2001) PROSPECTS, procedures for recommending optimal sustainable planning of European city transport systems: deliverable D1, report on work package 10: cities' decision-making requirements. Final report
34. Emberger G, Pfaffenbichler P, Jaensirisak S, Timms P (2008) "Ideal" decision-making processes for transport planning: a comparison between Europe and South East Asia. Transp Policy (Oxf) 15:341–349
35. Iannone F (2009) Analisi multicriteria per la classificazione di possibili interventi di potenziamento logistico dell'Area vasta pometina secondo le dimensioni ACIT. Working papers SIET 2009, ISSN 1973-3208
36. Skibniewski MJ, Chao L-C (1992) Evaluation of advanced construction technology with AHP method. J Constr Eng Manag 118:577–593
37. Hastak M, Halpin DW (2000) Assessment of life-cycle benefit-cost of composites in construction. J Compos Constr 4:103–111
38. Wankhade RL, Landage AB (2013) Non-destructive testing of concrete structures in Karad region. Procedia Eng 51:8–18
39. Kalamaras GS, Brino L, Carrieri G et al (2000) Application of multicriteria analysis to select the best highway alignment. Tunn Undergr Space Technol 15:415–420
40. Kayastha P, Dhital MR, De Smedt F (2013) Application of the analytical hierarchy process (AHP) for landslide susceptibility mapping: a case study from the Tinau watershed, west Nepal. Comput Geosci 52:398–408
41. Bitarafan M, Arefi SL, Zolfani SH, Mahmoudzadeh A (2013) Selecting the best design scenario of the smart structure of bridges for probably future earthquakes. Procedia Eng 57:193–199
42. Broniewicz E, Ogrodnik K (2020) Multi-criteria analysis of transport infrastructure projects. Transp Res D Transp Environ 83:102351
43. Marsden G, Reardon L (2017) Questions of governance: rethinking the study of transportation policy. Transp Res Part A Policy Pract 101:238–251. https://doi.org/10.1016/J.TRA.2017.05.008

Chapter 4
Participation in Territorial Governance: A Group Decision-Making Context

Abstract This chapter explores group decision-making in territorial governance, with a focus on stakeholder participation in transport infrastructure planning. It examines the challenges of reconciling individual and collective preferences, emphasising the complexities of multi-stakeholder decision processes. Using Social Choice Theory and Social Welfare Functions as a basis, the chapter analyses mechanisms for aggregating diverse preferences into collective decisions,addressing the conflicts that are inherent in infrastructure planning. Multi-Criteria Analysis (MCA) methods are discussed as tools to structure decision-making processes and facilitate consensus. Integrating political, economic, and social considerations, enhances this framework ability to ptomote participatory governance and ensure more balanced and inclusive decision-making in sustainable transport development.

Transport infrastructure provides the accessibility and mobility that cities need to thrive and expand. However, alongside economic growth imperatives and increasing awareness of sustainable development needs, social issues emerge, particularly inequality and the resultant conflicts stemming from inequitable wealth distribution. These challenges are convoluted with concerns regarding the limits of natural resource exploitation associated with urbanization. Such issues have become central to territorial governance and the planning of major transport initiatives.

Consequently, comprehensive evaluation makes mobility-related decision-making a complex task, requiring trade-offs among multiple quantitative and qualitative performance metrics: social, economic, and environmental. Therefore, multi-stakeholder participation is considered crucial for successful sustainable transport project planning. However, while stakeholder involvement is appropriate, it rarely produces a mutually acceptable solution due to diverse and conflicting interests. Inadequate management of stakeholder conflicts and controversies has frequently led to project failures [1, 2]. Thus, to reduce potential conflicts among stakeholders and their representatives in evaluating and ultimately selecting transport projects, establishing a methodology that enables collective preference formation, incorporating individual stakeholder perceptions, becomes paramount.

E. M. Bertolini, *Temporal Value in MCDM*, Lecture Notes in Intelligent Transportation and Infrastructure, https://doi.org/10.1007/978-3-031-95485-6_4

The rapid growth of urban populations has placed significant pressure on urban transportation systems. It is important to recognize that infrastructure serves as a catalyst for the socioeconomic development of a region; however, it can also entail substantial social, environmental, and economic costs, particularly for disadvantaged groups [3]. Consequently, the absence of effective public participation can become a major barrier to successful transportation planning, even in developing countries [4].

In Italy, for instance, public involvement in transport planning is primarily confined to environmental impact assessment, through which social impact is also evaluated. Generally, the absence of stakeholders beyond decision-makers make transport planning a rather subjective political process [5].

While attempting to avoid this, analysis complexity increases substantially due to the transition from a single decision-maker context to a group decision-making framework.

The challenge amplifies particularly when selecting preferred alternatives among diverse interest groups with different objectives, purposes, and criteria, thus generating potentially conflicting preferred solutions. Consequently, group decision-making processes must be based on multiple criteria and encompass diverse, interconnected fields.

Furthermore, the criteria themselves, not merely the alternatives, may conflict, generating additional intra-group tensions. Internal group decision dynamics are studied through various theoretical approaches, including preference analysis, utility theory, social choice theory, voting theory, general game theory, expert evaluation analysis, qualitative factor aggregation, and economic equilibrium theory. This diversity of approaches demonstrates that multiple-criteria group decision-making problems are highly varied and intricate.

4.1 The Group Decision-Making Challenge: Reconciling Individual and Collective Preferences

In group decision-making contexts, a fundamental challenge emerges when attempting to reconcile individual preferences (i) with the collective preference system (C) of which the individual is a member ($i \in$ C). Within society, where individuals naturally pursue their self-interest, there often exists a misalignment between personal and public interests. This misalignment makes determining societal preferences particularly complex, as it necessitates the coexistence of multiple, frequently conflicting individual interests.

The aggregation of individual preferences or ordinal rankings into a group preference or consensus ranking represents a classic Group Decision Making (GDM) problem. The analytical complexity intensifies significantly when transitioning from single-decision-maker frameworks to group decision contexts. This shift transforms the challenge from selecting a preferred alternative based on an individual's preference structure to resolving conflicts among diverse interest groups with distinct objectives, goals, and criteria, leading to divergent preferred solutions.

Group decision-making fundamentally involves the synthesis of diverse individual preferences regarding elements within a given set into a single collective or group preference. This process essentially represents a transformation from multiple individual datasets into a unified collective dataset.

Traditional preference analysis assumes rational agents who share the common goal of optimizing objective functions subject to specific constraints. This framework enables the analysis of preference aggregation as a specialized case of mathematical optimization. In economic contexts, for instance, this manifests in the assumption that consumers maximize utility within budgetary constraints, while firms maximize profits subject to technological and market environmental constraints. While this optimization-theory approach provides a logical conceptual framework for decision-making processes, it also offers a powerful and coherent analytical tool for choice analysis among alternatives.

4.2 Social Choice Theory: Mechanisms of Collective Decision-Making

Social Choice Theory (SCT) examines the systems and institutions that facilitate collective decision-making processes. This theoretical framework investigates the mechanisms through which choices affecting groups can be determined, focusing on how individual interests, preferences, and judgments can be aggregated into collective decisions. The ultimate outcome of such processes is inherently dependent on both the information available to participants and the rules established by the collective.

In capitalist democracies, social choices are predominantly determined through two fundamental mechanisms: voting systems, typically employed for political decisions, and market mechanisms, utilized for economic decisions. The voting method, along with decisions directly or indirectly based upon it, aims to capture agent preferences through majority-expressing balloting systems [6].

However, the complexity of social choice extends beyond these basic mechanisms. Several critical considerations warrant attention:

First, the aggregation of individual preferences into collective decisions faces fundamental challenges identified in Arrow's Impossibility Theorem, which demonstrates that no voting system can simultaneously satisfy all criteria for fair decision-making while maintaining logical consistency. This theoretical constraint has profound implications for democratic decision-making processes.

Second, the interaction between political and economic decision mechanisms creates a complex dynamic. Market forces and political processes often influence each other, leading to hybrid decision-making systems that combine elements of both mechanisms. This interaction becomes particularly evident in areas such as public goods provision, regulatory policy, and resource allocation.

Third, information asymmetries and strategic behavior significantly impact social choice outcomes. Individual actors may possess different levels of information or deliberately misrepresent their preferences to achieve desired outcomes, complicating the aggregation process.

Fourth, the institutional framework within which social choices occur plays a crucial role in determining outcomes. Formal and informal rules, cultural norms, and organizational structures all influence how individual preferences are transformed into collective decisions.

Understanding these complexities is essential for developing more effective and equitable decision-making systems in modern societies. The theoretical insights of Social Choice Theory continue to inform practical applications in public policy, corporate governance, and institutional design.

4.2.1 Voting Systems: Multi-criteria Decision Processes and Their Complexities

Voting represents a complex multi-criteria decision-making process that manifests both individual preference aggregation and collective choice challenges. When voters cast their ballots to select candidates or policies, they engage in a multifaceted evaluation process, considering various criteria such as reliability, integrity, competence, and ideological positioning—whether conservative, moderate, or liberal—along with stances on specific issues. While voters synthesize these criteria through an implicit value function (utility function), the actual vote tabulation process does not explicitly capture these nuanced evaluations.

Structural Dichotomy in Electoral Systems

A critical distinction exists between the voting and counting processes within electoral systems. The voting process involves the entire eligible electorate across various geographical jurisdictions, while the counting process is conducted by a select group of trained personnel under expert supervision and rigorous oversight. This dichotomy creates competing demands: the voting process must maintain simplicity and accessibility for the general electorate, while the counting process prioritizes accuracy and effectiveness [7].

Preference Aggregation Paradoxes

Consider the following illustrative scenario with an electorate of 60 voters choosing among candidates a, b, and c:

23 voters	a P c P b
19 voters	b P c P a

(continued)

(continued)

16 voters	c P b P a
2 voters	c P a P b

where P denotes "preferred over".

This scenario demonstrates how different counting methods can yield disparate outcomes:

- Under plurality voting: Candidate a wins with 23 votes (versus b's 19 and c's 18)
- Under runoff voting: Candidate b could win in a second round against a (obtaining 35 votes [19 + 16] versus a's 25 [23 + 2])
- Candidate C, through the preference system, receives the highest number of preferences compared to candidates A and B.

b P a	19 + 16 = 35	a P b	23 + 2 = 25
c P a	19 + 16 + 2 = 37	a P c	23
c P b	23 + 16 + 2 = 41	b P c	19

This is a clearly undesirable situation, as each individual would prefer the voting method that elects their preferred candidate. In another example, simple majorities may be intransitive in a scenario where A defeats B, B defeats C, and C defeats A. This outcome is known as the voting paradox, this paradox is referred to as the Condorcet effect.

4.2.1.1 The Condorcet Paradox: A Fundamental Challenge in Social Choice Theory

The "voting paradox" demonstrates a fundamental inconsistency in social choice processes based on majority rule: while individual voters may hold transitive preferences, collective decision-making can result in intransitive (cyclic) rankings among candidates. This occurs when majority preferences create a cycle where alternative x defeats y, y defeats z, and z defeats x, violating the assumption of transitivity in rational choice.

The complexity of voting procedures was first explored in sophisticated studies by Borda in 1770 [8], followed by Condorcet's seminal analysis of these inconsistencies in 1785 [9]. The Marquis de Condorcet identified and analyzed this paradox, which subsequently became known as the Condorcet effect. His work laid the foundation for understanding the analytical challenges inherent in majority-rule systems.

Consider a committee of three voters (I, II, and III) choosing among three alternatives (x, y, and z) with the following preference orderings:

Voter I	x P y P z
Voter II	y P z P x
Voter III	z P x P y

If the committee composed of voters I, II, and III agrees to rank preferences using a majority rule, it will rank X over Y, Y over Z, but Z over X. In this case, the committee would exhibit a circular preference among the alternatives, failing to establish a transitive ranking.

The Condorcet paradox's significance extends beyond its mathematical curiosity. It catalyzed extensive research in social choice theory, notably through studies by Garman, Kamien, Niemi, and Weisberg, revealing fundamental contradictions in simple majority rule applications. This work has profound implications for: democratic Theory and institutional Design.

Understanding the Condorcet paradox remains crucial for modern democratic theory and practice, particularly as societies grapple with increasingly complex collective decision-making challenges.

The Condorcet Effect: Analysis of Collective Preference Paradoxes

Consider an assembly faced with selecting among three alternatives (a, b, and c), where each voter maintains a coherent set of preferences. The possible a priori preference orderings can be categorized into six distinct patterns:

1. a P b P c
2. a P c P b
3. c P a P b
4. c P b P a
5. b P c P a
6. b P a P c

where P denotes "preferred over".

Let n represent the total number of voters, with n_i (i = 1, 2, …, 6) denoting the number of voters in each preference category. Following Condorcet's decomposition principle, collective opinion analysis proceeds by examining pairwise comparisons:

For a versus b:	
Prefer a over b	{i: a Pi b} = n1 + n2 + n3
Prefer b over a	{i: b Pi a} = n4 + n5 + n6
For a versus c:	
Prefer a over c	{i: a Pi c} = n1 + n2 + n6
Prefer c over a	{i: c Pi a} = n3 + n4 + n5
For b versus c:	
Prefer b over c	{i: b Pi c} = n1 + n5 + n6
Prefer c over b	{i: c Pi b} = n2 + n3 + n4

It is possible that the three collective judgments are inconsistent with each other. Inconsistency, condition for cyclical preferences emerges under two scenarios:

- a P b, b P c and c P a, or
- b P a, c P b and a P c.

The first case arises when the following three inequalities hold simultaneously:

- $n_1 + n_2 + n_3 > n_4 + n_5 + n_6$
- $n_1 + n_5 + n_6 > n_2 + n_3 + n_4$
- $n_3 + n_4 + n_5 > n_1 + n_2 + n_6$

Second cycle type (b P a, c P b, a P c) occurs when these inequalities are reversed.

Empirical research by Garman and Niemi has demonstrated that the probability of encountering the Condorcet paradox (no majority winner) is a function of both voter count and alternative count [10, 11]. Their findings, based on equiprobable ranking orders, reveal two key insights: the probability of cyclic majorities increases with the number of alternatives, in contrast the probability shows minimal sensitivity to voter count variations.

Arrow's formal treatment of the Condorcet paradox demonstrated its fundamental nature in social choice processes. This "impossibility" has become a central concern in social choice theory, spurring the development of various mitigation strategies and alternative voting procedures throughout the 20th century.

Garman and Niemi demonstrated that the probability of a winner without a majority is a function of the number of voters and alternatives [10, 11]. Both studies also assume the equal probability of all possible ranking orders. However, it emerges that as the number of alternatives increases, the likelihood of cyclic majorities rises, whereas the probability of cyclic majorities shows little sensitivity to the number of voters.

Contemporary Considerations

Modern voting systems face a range of new complexities that build upon classical challenges. The rise of digital voting technologies presents opportunities for improved efficiency but also raises concerns about security, verification, and accessibility. At the same time, strategic voting has become more prevalent, with voters casting ballots based on tactical considerations rather than genuine preferences, potentially distorting collective decision-making. Additionally, information asymmetry, influenced by media and information dissemination, further complicates how voter preferences are shaped and expressed. Finally, the issue of representation quality remains central, as different voting systems vary in their ability to accurately translate individual preferences into collective decisions, particularly in diverse and multicultural societies.

These contemporary challenges underscore fundamental tensions in democratic theory, particularly the trade-offs between simplicity and accuracy, majority rule and minority protection, and procedural fairness and substantive outcomes. Addressing these complexities is essential for the development of effective voting systems that uphold democratic legitimacy while accommodating practical constraints. As new technologies and evolving social dynamics continue to shape electoral processes, the study of voting systems remains a critical and evolving field.

4.2.1.2 Arrow's Impossibility Theorem: Fundamental Constraints in Social Choice Theory

Kenneth Arrow's seminal 1951 work, republished in 2012 [12], established a fundamental theorem in social choice theory. Arrow investigated whether it was possible to construct a collective choice method that satisfies certain rational conditions while transitioning from known individual preferences to a social decision model [6].

Consider a community needing to establish preference orderings among multiple alternatives, where each individual holds distinct preference rankings expressible through voting. The challenge lies in developing a procedure (such as a voting system), more broadly termed a social choice function, that transforms individual preferences into a coherent global ordering.

Arrow axiomatically demonstrated the impossibility of determining a social choice function among three or more alternatives that simultaneously satisfies five fundamental conditions:

Universality (Unrestricted Domain): The system must function independently of preference distributions and the range of available alternatives. It must accommodate all possible preference profiles.

Non-Dictatorship: The social choice function cannot merely reflect the preference ordering of a single individual or subset, but must consider the broader collective's preferences.

Transitivity: Requires logical consistency in preference ordering—if A is preferred to B and B to C, then A must be preferred to C.

Unanimity (Pareto Efficiency): When all individuals prefer one alternative to another, the social welfare function must reflect this unanimous preference.

Independence of Irrelevant Alternatives (IIA): The social choice for any pair of alternatives depends solely on individual preferences over that pair. Changes in preferences over other alternatives should not affect the choice between the pair.

The theorem demonstrates that when the voting population includes at least two individuals and three or more alternatives exist, no social choice function can simultaneously satisfy all five conditions. This finding has profound implications for democratic theory and practice, highlighting inherent limitations in democratic decision-making processes and suggesting necessary trade-offs in voting system design.

The application of game theory to voting systems emerges as a natural consequence of Arrow's theorem, indicating potential utilities in predicting voting outcomes while suggesting possible inefficiencies in voting equilibria. This theoretical framework questions the achievability of purely democratic decision-making and emphasizes the need for careful institutional design.

While Arrow's theorem identifies fundamental limitations, research continues in alternative frameworks, including the development of approximate or relaxed conditions and exploration of different axiom sets.

The theorem's implications for Pareto efficiency and social welfare optimization continue to influence contemporary discussions in social choice theory and practical policy design, suggesting that while perfect democratic decision-making systems may be theoretically impossible, understanding these constraints enables the development of more effective and equitable voting mechanisms.

4.2.2 Pareto's Theory: Foundations of Social Welfare and Economic Efficiency

Vilfredo Pareto, an Italian economist who lived between the latter half of the 19th century and the early decades of the 20th century, developed his theoretical framework within the neoclassical or marginalist school of economic thought that emerged around 1870. His contributions fundamentally shaped our understanding of social welfare and economic efficiency.

Pareto's theoretical foundation rests on methodological individualism—the conviction that any social aggregate comprises individual actors, necessitating that scientific analysis of society must begin with examining individual behavior. This approach recognizes that each individual can make their own evaluations while simultaneously serving as the judge of their own interests. Taking individual preferences and resource endowments as given, Pareto sought to analyze how society could optimally utilize its scarce resources to maximize collective welfare, defining efficiency as the ability to achieve predetermined objectives while maximizing available resource utilization.

The Pareto criterion rests on two fundamental premises: first, situations must be evaluated based on the welfare they provide to individuals; second, welfare comparisons often become impossible or arbitrary across different scenarios. To address these challenges, Pareto formulated two pivotal social choice criteria that continue to generate scholarly discourse and debate. These criteria apply to a collective C comprising n individuals, denoted as $C = \{1, 2, \ldots, n\}$.

The **first criterion**, known as the Weak Pareto Criterion, states: Given two states A and B, a collective C prefers state B to state A in the weak Pareto sense if and only if all individuals strictly prefer B to A. Using formal notation, with individual preferences denoted in parentheses and group preferences in square brackets, this applies to any two states A and B, and n individuals, $i \in \{1, 2, \ldots, n\}$.

$$[\mathrm{B} > \mathrm{A}]_{\mathrm{P}} \Leftrightarrow \forall i \in \{1, 2, \ldots, n\} \text{ results } (\mathrm{B} > \mathrm{A})_{\mathrm{i}} \tag{4.1}$$

where A and B are two states of the world; $\{1, 2, \ldots, n\}$ represent the individuals.

The **second criterion**, the Strong Pareto Criterion, extends this concept: Given two states A and B, a collective C prefers state B to state A in the strong Pareto sense if and only if some individuals (potentially just one) strictly prefer B to A, while no one strictly prefers A to B. Under these conditions, state B is deemed superior to state A (or B dominates A) in the strong Pareto sense, and a transition from A to B constitutes a Pareto improvement.

$$[\mathrm{B} > \mathrm{A}]_{\mathrm{P}} \Leftrightarrow \forall i \in \{1, 2, \ldots, \mathrm{n}\}, \text{ so}$$

$$[\mathrm{B} > \mathrm{A}]_{\mathrm{i}}\mathrm{e}\ \forall j \neq i \in \{1, 2, \ldots, \mathrm{n}\}(\mathrm{A} > /\mathrm{B})_{\mathrm{j}} \tag{4.2}$$

where A and B are two states of the world; {1, 2,..., n} represent the individuals.

These criteria have profound implications for economic policy and social choice theory. They provide a framework for evaluating social welfare changes while respecting individual preferences, though they also reveal the inherent challenges in making social welfare comparisons. Contemporary applications extend beyond economics to areas such as public policy, environmental regulation, and institutional design, where policymakers must balance competing interests while striving for efficient resource allocation.

The enduring relevance of Pareto's theoretical framework lies in its ability to provide clear criteria for evaluating social welfare improvements while acknowledging the complexity of interpersonal welfare comparisons. However, it's important to recognize that Pareto efficiency, while necessary for optimal resource allocation, may not always align with other social objectives such as equity or fairness, highlighting the ongoing tension between efficiency and distribution in economic policy.

The Pareto criteria provide a framework for comparing two societal states or situations (A and B) from a collective societal perspective. These criteria, and the resulting social ordering, are fundamentally constructed from individual preferences, leading to an indirect ordering system that notably abstains from explicit equity considerations. This framework's limitation becomes apparent when confronted with questions regarding the distribution of advantages or the relative improvements experienced by different socioeconomic groups, whether more or less privileged. Such questions remain unanswered due to the postulate of ethical individualism that underlies Pareto's approach.

The theoretical foundation rests on the premise that each individual is motivated by their own objectives (preferences) and operates within constraints determined by resource availability, technological capabilities, and various impediments. The primary analytical objective is to delineate the conditions that characterize the final equilibrium between preferences and constraints.

While the Pareto criterion serves as a comparative tool between two alternatives, the concept of Pareto optimality aims to identify the preferred alternative(s) among all available options. A situation is considered Pareto optimal when no alternative exists that would improve at least one individual's welfare without diminishing that of others. This concept embodies an efficiency condition wherein, given fixed resources, the possibility exists to enhance the situation of one or more individuals without adversely affecting others (and potentially even improving their condition). This optimization occurs through logical actions where means are directly aligned with desired ends.

4.2.2.1 Pareto Optimality: A Framework for Social Welfare Analysis

A Pareto optimal state, or Pareto efficiency, represents a condition in social welfare theory where no further reallocation of resources can improve one individual's situation without adversely affecting another's welfare. More formally, given a collective C, a state of the world SC is considered Pareto efficient when no Pareto improvement is possible—that is, when any deviation from this state cannot enhance at least one individual's welfare without diminishing that of another.

This fundamental concept in welfare economics reveals a crucial limitation in social choice theory: the impossibility of definitively determining preference between two situations where different individuals experience opposing welfare effects. For instance, if Situation A benefits certain members of society while disadvantaging others compared to Situation B, Pareto criteria alone cannot establish which state is socially preferable. This limitation underscores the persistent challenge of aggregating individual preferences into a coherent collective preference system.

The implications of Pareto optimality extend beyond theoretical considerations to practical policy applications:

First, multiple Pareto optimal states may exist within a given economic system, each representing different distributional outcomes. While all these states are efficient in the Pareto sense, they may vary significantly in terms of equity and social justice.

Second, achieving Pareto efficiency does not necessarily result in socially desirable outcomes. A situation could be Pareto optimal while still maintaining significant inequality or other socially undesirable characteristics.

Third, the concept's practical application in policy-making is limited by its binary nature—it can identify whether a change creates winners and losers but cannot weigh the magnitude of gains against losses.

Understanding these limitations and extensions of Pareto optimality is crucial for policy makers and economists working to design and implement effective social welfare policies while balancing efficiency and equity considerations.

4.2.3 The Social Choice Function: Mechanisms of Collective Decision-Making

The complexity of social choice mechanisms extends deeply into the analysis of how collective decisions can be implemented within real-world contexts, considering both practical constraints and the resulting interactions among individuals. This investigation requires a rigorous mathematical framework to formalize the relationship between individual preferences and collective outcomes.

Mathematical Formalization

A social choice correspondence can be formally defined as follows.

Let N represent a set of n agents and A represent a set of alternatives. If we denote P as the set of weak preference orderings over A, with Q being a subset of P, then a Social Choice Correspondence (SCC) can be expressed as:

$$\mathcal{C} : \mathcal{Q}^n \rightrightarrows \mathcal{A} \tag{4.3}$$

where C represents the social choice correspondence function, Q denotes the series of preferences over A, A represents the set of alternatives, n indicates the number of agents in set N.

When the correspondence yields only single elements, it becomes a Social Choice Function (SCF), expressed as:

$$\mathcal{F} : \mathcal{Q}^n \rightrightarrows \mathcal{A} \tag{4.4}$$

where F represents the social choice function, and other variables maintain their previous definition.

Theoretical Implications

The social choice function essentially operates as a mechanism that assigns a preferred alternative to every possible combination of voter preferences. This mathematical formalization captures the fundamental challenge of aggregating individual preferences into collective decisions.

The generality of these results becomes particularly significant when considering Arrow's Impossibility Theorem. For the findings to maintain broad applicability—that is, to be valid for societies whose members possess the most general possible individual preference systems—the preference set Q must approximate P as closely as possible.

Contemporary Applications and Extensions

This theoretical framework has several important implications for modern institutional design:

- it provides a foundation for analyzing voting systems and other collective decision-making mechanisms, allowing for systematic evaluation of their properties and limitations.
- it enables the study of strategic behavior in collective choice situations, as individuals may attempt to manipulate outcomes through strategic preference revelation.

- it informs the design of practical decision-making procedures in various contexts, from public policy to corporate governance.

The challenge lies in bridging the gap between theoretical elegance and practical implementation, particularly in contexts where preference aggregation must account for real-world constraints and strategic behavior. Understanding these limitations while maintaining mathematical rigor remains crucial for developing effective social choice mechanisms.

This framework continues to evolve as new computational tools and theoretical insights emerge, offering fresh perspectives on the fundamental challenge of converting individual preferences into collective decisions in a manner that satisfies both theoretical requirements and practical constraints.

4.3 Social Welfare Functions: Bridging Individual Preferences and Collective Well-Being

The inability to derive a complete social ordering based solely on the Pareto criterion, even when supplemented with additional conditions, demonstrates the limitations of analyzing efficiency in isolation from equity considerations. This limitation necessitates evaluating collective choice as an aggregation of individual preferences. A social welfare function (SWF) serves as a mechanism for mapping individual preference orderings into possible societal preference orderings.

The mathematical formalization of this aggregation can be expressed as:

$$W = f(u_1 + u_2 + \cdots + u_n) \tag{4.5}$$

where W represents the general social welfare function and u_i denotes the utility functions of n individuals comprising society.

The social welfare function W associates a social state A with the quantity W(A), representing the collective welfare enjoyed by society in state A. Intuitively, SWFs align with the Pareto principle, suggesting that social welfare increases (or at least remains constant) when one or more individual utilities increase while others remain unchanged, expressed as:

$$\frac{\partial W}{\partial u_i} \geq 0, \forall i \in \{0, \ldots, n\} \tag{4.6}$$

where W is the general social welfare function, u_i is the utility of individual i, and i represents a generic individual.

First, while numerous social welfare functions are theoretically possible, only some may be deemed acceptable by society. Acceptable functions should satisfy principles of fairness and ethical admissibility, given their social implications.

Second, to achieve complete social orderings, individual preferences must be comparable, particularly in situations where changes benefit some while disadvantaging others. This necessity leads to discussions of distributive and social justice principles that simultaneously accept ethical individualism and interpersonal utility comparability.

Third, implementing these functions requires accepting that social welfare should reflect individual preferences regarding various states while acknowledging the possibility of interpersonal comparisons—a necessary condition for achieving complete orderings.

The principal categories of social welfare functions include:

- Utilitarian SWFs
- Bernoulli-Nash SWFs
- Rawlsian SWFs
- Bergson-Samuelson SWFs.

Each approach offers distinct perspectives on aggregating individual preferences into collective welfare. The selection of an appropriate mathematical operator (such as summation) for synthesizing individual preferences into a single social preference becomes crucial.

Contemporary applications of social welfare functions extend beyond theoretical economics to practical policy decisions, including:

- Income distribution policies
- Public goods provision
- Environmental regulation
- Healthcare resource allocation.

Understanding these functions' properties and limitations remains essential for designing effective social policies that balance efficiency with equity considerations while respecting both individual preferences and collective welfare objectives.

4.3.1 The Utilitarian Social Welfare Function: Benthamian Approach to Collective Welfare

The utilitarian social welfare function (USWF) represents the most straightforward approach to aggregating individual utilities into a measure of social welfare. In its basic form, the social welfare function is constructed as the sum of individual utility functions. When individual utilities are comparable and each person is assigned equal (unitary) weight, the social utility of a state A, denoted as W(A), can be expressed as:

$$W(A) = \sum_{i=1}^{n} u_i \tag{4.7}$$

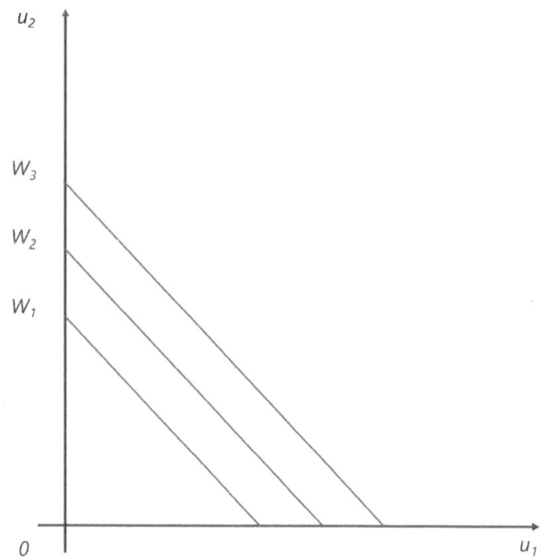

Fig. 4.1 Social indifference curves of the utilitarian welfare function

where W(A) represents the social utility of state A, u_i denotes individual utility functions for n members of society.

This framework can be visualized through indifference curves, known in this context as social indifference curves or social iso-well-being curves. These appear as parallel lines with negative 45-degree slopes, denoted as Wi, where W1 < W2 < W3. Along each line, the sum of utilities remains constant. These iso-well-being curves are plotted on a plane defined by the utilities (u_1, u_2) of two considered individuals (Fig. 4.1).

A more sophisticated version of the utilitarian approach emerges through the generalized utilitarian social welfare function, which assigns different non-negative weights (a_i) to individuals:

$$W(A) = \sum_{i=1}^{n} a_i u_i \tag{4.8}$$

where W(A) remains the social utility of state A, u_i represents individual utility functions, a_i denotes non-negative weights assigned to each individual.

This social welfare relationship can be graphically represented through a map of social indifference curves, which manifest as parallel lines with negative slopes proportional to the weights assigned to different individuals. When considering a simplified two-person case with utilities u_1 and u_2, these indifference curves take the form of linear functions expressed as $a_1u_1 + a_2u_2 = k$, where the slope, measured by the tangent of angle β, equals-a_1/a_2 (Fig. 4.2).

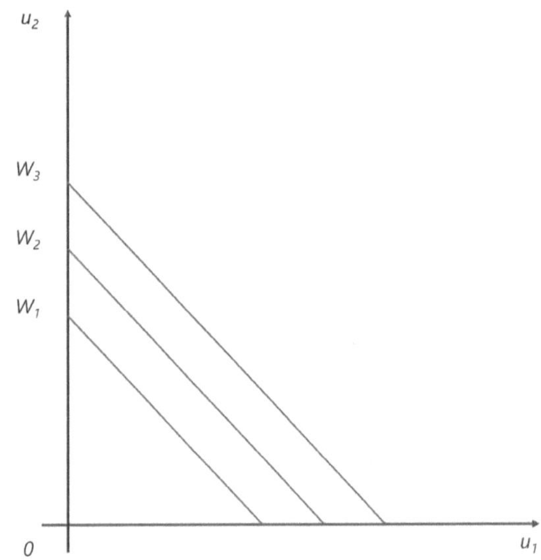

Fig. 4.2 Social indifference curves of the generalized utilitarian SWF. *Image Source* Author's elaboration

The utilitarian Social Welfare Functions (SWFs) employ summation as their fundamental mathematical operator. Their implementation necessitates two critical conditions:

- Cardinal measurability of individual utilities: welfare must be quantifiable through measurable variables, such as monetary income.
- Comparability of individual welfare states: each individual must carry equal weight in the aggregate calculation.

Utilitarian SWFs, by virtue of their additive nature, maintain neutrality toward utility inequalities. This characteristic allows for welfare improvements when utility gains for some individuals outweigh utility losses for others, thereby transcending the limitations of strict Pareto efficiency criteria. This framework permits utility losses for certain individuals without mandating universal improvement or maintenance of utility levels, thus departing from egalitarian principles. Nevertheless, the utilitarian function maintains consistency with the strong Pareto criterion, as improvements in at least one individual's utility, absent negative effects on others, enhance overall social welfare.

The utilitarian approach, however, faces several notable criticisms, such as: It assumes cardinal utility measurement and interpersonal utility comparisons, which many economists consider problematic.; the simple summation of utilities may not adequately account for distributional concerns or equity considerations; and the assignment of weights in the generalized form introduces potential subjectivity and political considerations into what appears to be an objective framework.

Contemporary applications of utilitarian social welfare functions often appear in: public policy analysis, cost–benefit assessment, resource allocation decisions, environmental policy, healthcare resource distribution.

Understanding these strengths and limitations remains crucial for policymakers and analysts employing utilitarian frameworks in practical decision-making contexts. Modern approaches often combine utilitarian considerations with other social welfare criteria to create more comprehensive evaluation frameworks that better reflect society's multiple objectives and ethical considerations.

This unique property of utilitarian SWFs—allowing for compensatory trade-offs between individual utilities while remaining compatible with Pareto efficiency—highlights their distinctive role in welfare economics. The framework's ability to accommodate utility redistributions while maintaining mathematical tractability makes it a powerful tool for social welfare analysis, despite its potential tensions with strict egalitarian principles.

4.3.2 Rawls' Social Welfare Function

Rawls's Social Welfare Function (SWF) evaluates societal welfare through the fundamental lens of fairness. His conception of justice as fairness posits that a society's core institutions can be considered equitable only when they guarantee three essential principles:

- Equal civil and political rights and liberties for all citizens
- Equality of opportunity, manifested as the absence of unjustified discrimination in access to social roles and career advancement
- Resource distribution governed by the principle that social welfare increases only when the position of the least advantaged members of society improves (max–min principle)

This theoretical framework operates under the maximin hypothesis: societal welfare is fundamentally determined by the welfare of its poorest member and can only increase when the condition of the least advantaged improves. Mathematically, this can be expressed as:

$$W = \min(u_1, u_2, \ldots, u_n) \tag{4.9}$$

where W represents the social utility state and u_i represents the utility functions of the n individuals comprising society.

In Rawls's SWF formulation, social welfare is directly equated with the satisfaction level of the least advantaged individual. This principle establishes that societal welfare is maximized only when the minimum satisfaction level increases, hence the designation "maximin SWF." The construction of Rawls's SWF requires two key informational prerequisites:

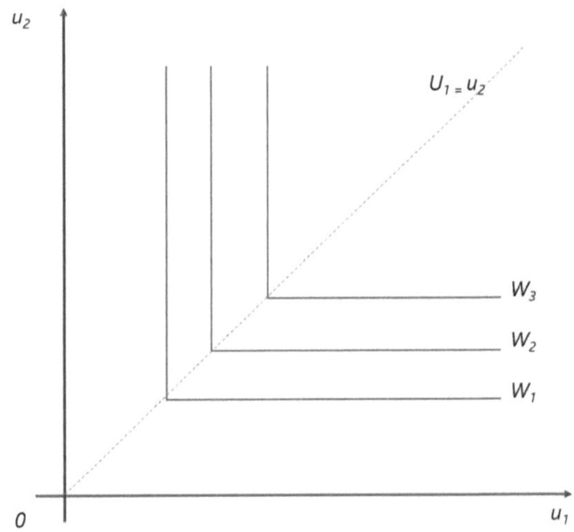

Fig. 4.3 Social indifference curves of the Rawlsian SWF. *Image Source* Author's elaboration

1. Ordinal measurability.
2. Comparability of individual utility levels.

In a two-individual scenario with utilities u_1 and u_2, Rawls's SWF can be graphically represented through L-shaped iso-welfare curves, characterized by vertices along the first quadrant's bisector with sides parallel to the axes. This distinctive L-shape reveals a crucial property: increases in one individual's utility do not enhance social welfare (which remains at levels W_1, W_2, or W_3) if the least advantaged individual's utility remains constant. This indicates perfect complementarity between individual satisfactions, rather than substitutability as found in other SWF formulations (Fig. 4.3).

Rawls's SWF uniquely synthesizes egalitarian and utilitarian elements. The egalitarian aspect manifests in the priority given to the welfare of the least advantaged in assessing societal well-being. The utilitarian dimension emerges in the function's indifference to inequalities affecting the broader population once the welfare of the poorest individual is secured. This theoretical framework thus provides a sophisticated approach to balancing competing social welfare considerations while maintaining a strong commitment to protecting society's most vulnerable members.

4.3.3 The Bernoulli-Nash Social Welfare Function

The Bernoulli-Nash Social Welfare Function (SWF) distinguishes itself by employing multiplication as its fundamental mathematical operator. This framework manifests in two forms: the simple Bernoulli-Nash SWF and its generalized variant.

The simple form is expressed as:

$$W = u_1 * u_2 * \cdots * u_n = \prod_1^n u_i \tag{4.10}$$

where W represents the social utility state and u_i denotes the utility functions of the n individuals comprising society.

The generalized Bernoulli-Nash SWF introduces weighted exponents (a_i) to individual utilities in the multiplicative product:

$$W = \prod_1^n (u_i)^{a_i} \tag{4.11}$$

where W maintains its definition as social utility state, u_i represents individual utility functions, a_i denotes non-negative weights assigned to each individual.

In a two-person society with utilities u_1 and u_2, the Bernoulli-Nash SWF generates hyperbolic iso-welfare curves in the utility plane. Notably, in the simple form, these curves manifest as equilateral hyperbolas (Fig. 4.4).

The multiplicative operator in the Bernoulli-Nash formulation yields an important property: social welfare increases with greater equity in utility distribution among individuals. This characteristic imbues the Bernoulli-Nash SWF with stronger egalitarian tendencies than the utilitarian function, demonstrating enhanced consideration for the welfare of disadvantaged individuals. However, it differs fundamentally from the Rawlsian function in that it permits compensation of losses to the poor

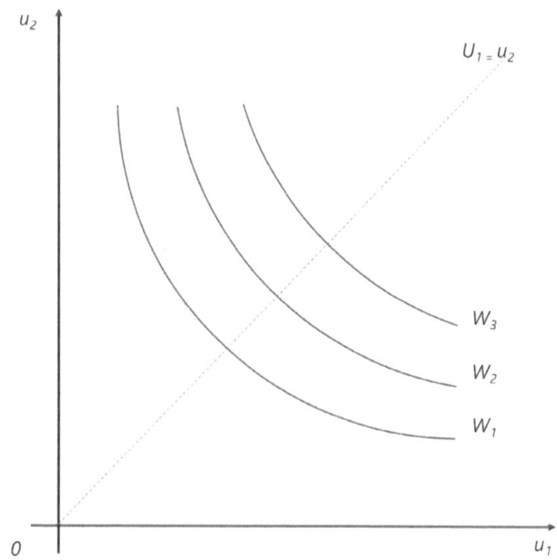

Fig. 4.4 Social indifference curves of the Bernoulli-Nash SWF. *Image Source* Author's elaboration

through substantial gains to the wealthy. This theoretical framework thus occupies a middle ground between strict egalitarianism and pure utilitarianism, offering a nuanced approach to social welfare evaluation.

This mathematical structure creates a sophisticated balance between equity considerations and efficiency improvements, making it particularly valuable for analyzing policy interventions that involve complex trade-offs between different societal groups. The function's hyperbolic properties ensure that extreme inequality is penalized while maintaining sensitivity to Pareto improvements across the entire welfare distribution.

4.3.4 The Isoelastic Social Welfare Function

The Isoelastic Social Welfare Function serves as a generalized framework that encompasses previously discussed welfare functions as special cases. Its mathematical formulation is expressed as:

$$\mathrm{W} = \frac{\sum_{i=1}^{n} \alpha_i (u_i)^{1-\rho}}{1-\rho} \tag{4.12}$$

where W represents the social utility state, u_i denotes individual utility functions for the n members of society, α_i represents non-negative weights assigned to individuals, and ρ is a non-negative constant parameter linked to the elasticity of substitution (σ) of social indifference curves through the relationship $\sigma = 1/\rho$.

The marginal social welfare for individual i can be derived as:

$$\frac{\partial W}{\partial u_i} = \alpha_i (u_i)^{-\rho} \tag{4.13}$$

where W is the social utility of the state, u_i is the utility of individual I, α_i is the non-negative weight assigned to individual i, and ρ is the constant parameter greater than 0 related to the elasticity of substitution of the social indifference curves (σ), where $\sigma = 1/\rho$.

This formulation's remarkable flexibility allows it to reproduce various established welfare functions through parameter adjustments:

1. When $\rho = 0$, it reduces to the utilitarian function ($\sigma = +\infty$)
2. As ρ approaches 1, it converges to the Bernoulli-Nash function ($\sigma = 1$)
3. As ρ approaches infinity, it approximates the Rawlsian function ($\sigma = 0$) (Fig. 4.5).

The isoelastic function's versatility extends beyond these special cases, as varying values of ρ generate intermediate curvatures of social indifference curves. This property enables the function to model how society evaluates trade-offs between utility

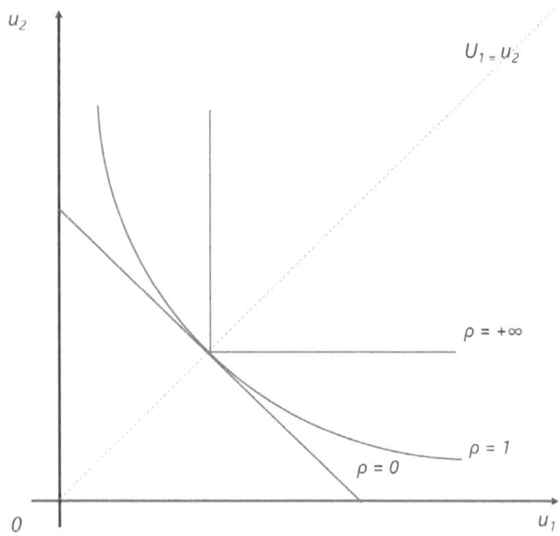

Fig. 4.5 Social indifference curves of the isoelastic SWF as ρ changes. *Image Source* Author's elaboration

levels across different groups or individuals with exceptional precision. The continuous spectrum of possible curvatures provides a sophisticated analytical tool for examining societal preferences regarding inequality and redistribution.

This mathematical framework thus offers a comprehensive approach to social welfare analysis, capable of representing diverse societal attitudes toward equality and efficiency through a single parametric specification. Its ability to nest multiple established welfare functions within a unified structure makes it particularly valuable for comparative analysis and policy evaluation across different social welfare paradigms.

4.3.5 *Comparative Analysis of Social Welfare Criteria*

Let us examine three distinct social welfare evaluation methods through their comparative application to three world states (A, B, C):

1. The Pareto Criterion
2. The Utilitarian Social Welfare Function (SWF)
3. The Rawlsian Social Welfare Function.

Consider three world states with corresponding utility indices for three individuals and their aggregate utilities. Through this comparative lens, several critical observations emerge (Table 4.1).

Under Pareto efficiency analysis, states A, B, and C prove incomparable as they all represent Pareto optimal points. The utilitarian criterion, focusing on aggregate

Table 4.1 Comparison between the Pareto criterion, the utilitarian SWF, and the Rawlsian SWF, where A, B, and C represent the states of the world, and u1, u2, and u3 are the utility indices of the three individuals considered

Individual utility	World state A	World state B	World state C
u_1	50	80	80
u_2	70	80	60
u_3	90	50	70
$\sum_{i=1}^{3} u_i$	210	210	210

Source Author's elaboration

utility, renders these states equivalent [A $\approx$ B $\approx$ C] due to identical utility sums. However, the Rawlsian perspective favors state C, as it maximizes the utility of the least advantaged individual relative to states A and B.

This analysis can be generalized by defining SC as the set of possible social states for collectivity C, and P, U, and R as the improvement sets according to Pareto, utilitarian, and Rawlsian criteria respectively. This formalization yields three fundamental relationships:

1. Pareto Dominance and Utilitarianism:
   - $\forall$ A, B $\in$ SC, A > P B $\Longrightarrow$ A > U B, but A > U B $\Longrightarrow$// A > P B
     Therefore, P $\subseteq$ U
2. Pareto Dominance and Rawlsian Criterion:
   - $\forall$ A, B $\in$ SC, A > P B $\Longrightarrow$// A > R B and A > R B $\Longrightarrow$// A > P B
     However, P $\cap$ R $\neq \emptyset$
3. Utilitarian and Rawlsian Criteria:
   - $\forall$ A, B $\in$ SC, A > U B $\Longrightarrow$// A > R B and A > R B $\Longrightarrow$// A > U B (Fig. 4.6).

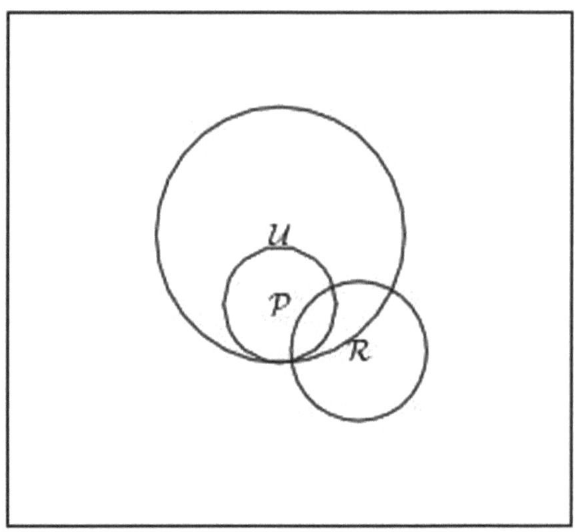

Fig. 4.6 Graphical representation of the improvement sets

Nevertheless, $U \cap R \neq \emptyset$

This formalization reveals the complex interrelationships between different welfare criteria, highlighting that while there are areas of overlap between these approaches, each captures distinct aspects of social welfare. The Pareto criterion's improvements constitute a subset of utilitarian improvements, while both maintain non-empty intersections with Rawlsian improvements without fully containing or being contained by them. This theoretical framework provides a sophisticated basis for understanding the complementary yet distinct nature of these welfare evaluation methods.

The analysis demonstrates that these criteria offer different perspectives on social welfare evaluation, each emphasizing particular aspects of societal well-being: efficiency (Pareto), aggregate welfare (utilitarian), or the condition of the least advantaged (Rawlsian). Their systematic comparison provides valuable insights for policy analysis and social choice theory.

4.3.6 Concluding Analysis

Contemporary research on group decision-making processes predominantly focuses on ordinal rankings of individual preferences across alternative sets, rather than cardinal utility analysis. Within this framework, a social ordering derived from individual preference sets represents an aggregation of alternatives that accounts for individual preference rankings while necessarily excluding the intensity of these preferences. Arrow's innovative application of symbolic logic methods has fundamentally shaped the entire conceptual foundation of social choice theory. His impossibility theorem, through the establishment of seemingly reasonable axioms, demonstrates the inherent impossibility of constructing an ideally rational aggregation mechanism for social preferences [7].

This theoretical constraint has profound implications for our understanding of collective decision-making processes. The emphasis on ordinal rather than cardinal preferences reflects both practical limitations in measuring preference intensities and theoretical challenges in interpersonal utility comparisons. Arrow's contribution extends beyond the specific impossibility result, introducing a rigorous mathematical framework that continues to influence contemporary approaches to social choice analysis.

The theorem's enduring significance lies not merely in its negative result, but in its illumination of the fundamental tensions between individual preferences and collective rationality. This understanding has shaped subsequent developments in welfare economics, voting theory, and public choice analysis, highlighting the inherent complexities in aggregating individual preferences into coherent social decisions.

4.4 Analysis and Multicriteria Methods Used in Group Decision Making

The Analysis and Evaluation of Territorial Governance Policies can be conducted through either non-participatory methods (led by analysts) or participatory approaches. This fundamental dichotomy in methodological framework presents distinct advantages and challenges for policy evaluation.

In non-participatory evaluations, technical experts independently conduct analyses using one or multiple analytical models. These specialists collect, process, and interpret data utilizing various decision support tools and methodologies, ultimately providing recommendations to decision-makers. This approach is predicated on the premise that trained specialists are better equipped to support complex and critical decision-making processes.

Conversely, participatory techniques employ a more collaborative decision-making style that directly engages multiple stakeholders. This inclusive approach enables analysts and decision-makers to consider previously overlooked perspectives, excluded possibilities, and neglected issues in their comprehensive analysis.

The selection of the most appropriate approach is contingent upon both the nature of the problem under examination and the available resources. Theoretically, an analyst-driven approach without stakeholder input may be more suitable for addressing purely technical problems characterized by relatively low levels of uncertainty and ambiguity. In contrast, more complex and uncertain policy issues affecting broader society may be better addressed through participatory processes, despite their longer duration and higher costs. This latter approach aims to ensure adequate representation of diverse viewpoints regarding the decision-making situation [13–16].

These approaches represent opposite ends of a methodological spectrum. Between the purely technical approach and the fully participatory model that involves all stakeholders through public debates, there exists a range of intermediate approaches facilitating various modes of interaction between technical experts and decision-makers.

While the Multi-Criteria Analysis (MCA) model was initially employed in a non-participatory manner, research over the past three decades has evolved beyond this specialist model. This evolution has been driven by increasing demands for public participation in planning and decision-making processes [16–20].

Methodologies combining participatory and deliberative procedures have become increasingly prevalent across various fields of planning and policy-making [21–28]. This evolution has led to the development of numerous multi-actor and multi-criteria approaches specifically designed for the transport sector.

Participatory MCA methods typically involve two primary categories of actors. The first consists of a research group comprising specialized analysts and consultants who manage the process while maintaining, to the greatest extent possible, an independent and comprehensive view of the issue at hand. This group ultimately presents the analysis results to decision-makers. The second category encompasses group decision-making participants, typically stakeholders directly involved in the problem, and in some instances, academics and experts who contribute scientific perspectives to the analysis. These participants may engage in the multi-criteria and multi-actor exercise either as individuals or as representatives of organized groups, such as local community organizations, landowners, business associations, or environmental experts.

From an operational perspective, participatory MCA methodologies closely mirror analyst-led MCA processes, encompassing several key phases that may occur in varying sequences and modes. These phases include option development, identification of objectives and criteria for testing options, criteria weighting, and scoring of option impacts against different criteria. However, participatory MCA techniques distinctively enable group decision-making participants to contribute to identifying key elements of the multi-criteria framework, including options, objectives and criteria, weights, and scores.

Qui di seguito si presenteranno alcuni metodi partecipativi di MCA proposti per il settore dei trasporti e le loro caratteristiche chiave.

Early contributions to participatory MCA in the transport sector emerged in the late 1960s when Hill [29, 30] introduced a simple additive weighting model combined with Cost–Benefit Analysis (CBA) elements. While these applications were predominantly implemented non-participatively in the United Kingdom, they did involve stakeholder groups, though without specifying group sizes or engagement techniques. Individual preferences were processed through mathematical aggregation of weights provided by various stakeholder groups.

Similarly, scholars including Leleur, Barfod, Jensen, and Salling [31–37] developed customized decision support systems and evaluation frameworks. These frameworks combined weighted additive model-based MCA, CBA, and various sensitivity analyses. While their approach did not specify participant types or numbers, it allowed stakeholders to influence objectives/criteria and weights through workshops and decision conferences, facilitating agreement on objectives and weightings.

This evolution in methodology reflects a growing recognition of the importance of stakeholder engagement in transport planning and decision-making processes, while maintaining analytical rigor through structured evaluation frameworks.

Notable contributions to multi-criteria and multi-actor analyses have been made by Macharis et al. [38, 39], and Macharis and Nijkamp [40, 41], with numerous applications particularly in Belgium. Their methodology incorporates AHP and PROMETHEE techniques, along with other MCA methods. While the practical implications remain somewhat unclear, their approach typically involves a limited number of key stakeholder groups (usually 3–6), though the number of stakeholders per group is not specified. Stakeholder perspectives, gathered through interviews, workshops, and surveys, are maintained separately and subsequently aggregated mathematically.

Dimitriou et al. [42, 43] and Ward et al. [43] developed a multi-criteria analysis utilizing a simple additive weighting model, primarily implemented in theoretical role-playing exercises. Their approach allows participants to partially influence options and objectives/criteria while directly determining scoring. Through workshop sessions, different group perspectives are kept distinct and used to explore potential interest convergence among various actors.

The ANP/InViTO approach (participatory MCA combined with data visualization tools) was developed by Lami and colleagues within the Interreg IVB NWE project framework. This method employs an Analytic Network Process (ANP) model, a variation of AHP, engaging several dozen participants including citizen representatives, decision-makers, and experts. Participants can influence options, criteria priorities (weights), and local priorities of each option relative to each criterion (scores) through workshops and questionnaires, supported by data visualization tools.

De Brucker et al. developed a two-stage MCA method using the AHP technique, involving a select number of key stakeholder groups (though specific group sizes remain undefined). Through workshops, stakeholders can influence objectives/criteria and weights, with different group perspectives maintained separately to explore potential interest convergence among various actors [44].

However, this overview reveals significant uncertainties regarding the practical implementation of these techniques. It remains unclear whether they represent purely academic proposals or have been successfully applied in practice, and how they might adapt to conventional planning procedures, evaluation methods, and public inquiry [45, 46].

The methodological adaptations of MCA to group decision-making have primarily occurred in three main domains [45]:

- Identification, classification, and selection of group decision-making participants
- Stakeholder (and expert) engagement in analysis and group process management
- Collection, processing, and integration of group decision-making participants' preferences within the multi-criteria framework.

This evolution in methodology highlights the increasing complexity of stakeholder engagement in decision-making processes while raising important questions about practical implementation and integration with existing planning frameworks.

Each domain of participatory Multi-Criteria Analysis (MCA) presents distinct methodological challenges and critical considerations. Regarding participant identification and selection, with notable exceptions [17, 43], the literature on participatory MCA remains largely ambiguous. This ambiguity creates a fundamental tension in the methodology's implementation.

Proponents of multi-actor multi-criteria methods [40], echoing discourse on deliberative democracy [47–50], emphasize that ideally, all stakeholders affected by the issue under consideration should be involved or represented in an MCA exercise, with no perspective excluded a priori. However, practical constraints related to process manageability and efficiency significantly limit the number of participants in group decision-making processes. A comprehensive review of participatory MCA techniques [45] revealed that such processes rarely involve more than 30 participants in total. This limitation raises significant concerns, as evaluation exercises addressing major project and policy issues with such restricted participation fail to meet statistical representativeness requirements and may paradoxically represent a regression in terms of democratic principles and equity [45, 51].

Stakeholder engagement can be facilitated through various participatory techniques, ranging from basic structured interviews and questionnaires to in-depth group discussions. Each approach presents distinct advantages and limitations. Individual interviews, for instance, while preventing direct interaction and discussion between different participants and groups, enable more thorough investigation of individual participants' interests and priorities while avoiding groupthink phenomena [46].

In an effort to facilitate processes and generate productive discussions, some methodological approaches [52–54] extensively utilize specialized software and visualization tools. These technological solutions aim to bridge the gap between theoretical ideals of comprehensive stakeholder participation and practical limitations of group decision-making processes.

This analysis highlights the ongoing challenge of balancing inclusive participation with practical implementation constraints in participatory MCA, suggesting the need for continued methodological innovation to address these tensions.

References

1. Olander S (2007) Stakeholder impact analysis in construction project management. Constr Manag Econ 25:277–287
2. Li THY, Ng ST, Skitmore M (2013) Evaluating stakeholder satisfaction during public participation in major infrastructure and construction projects: a fuzzy approach. Autom Constr 29:123–135
3. UN (United Nations) (2014) World urbanization prospects. New York
4. Shan C, Yai T (2011) Public involvement requirements for infrastructure planning in China. Habitat Int 35:158–166
5. Jones S, Tefe M, Appiah-Opoku S (2013) Proposed framework for sustainability screening of urban transport projects in developing countries: a case study of Accra, Ghana. Transp Res Part A Policy Pract 49:21–34
6. Arrow KJ, Vittadini G, Graziola G (2003) Scelte sociali e valori individuali. Etas
7. Hwang C-L, Lin M-J (2012) Group decision making under multiple criteria: methods and applications. Springer Science & Business Media
8. Borda JC (1770) Eclaircissement sur les méthodes de trouver ler courbes qui jouissent de quelque propiété du maximum ou du minimum. Mém Acad Roy Sci Paris 551–565
9. De Condorcet N (2014) Essai sur l'application de l'analyse à la probabilité des décisions rendues à la pluralité des voix. Cambridge University Press
10. Garman MB, Kamien MI (1968) The paradox of voting: probability calculations. Behav Sci 13:306–316
11. Niemi RG, Weisberg HF (1968) A mathematical solution for the probability of the paradox of voting. Behav Sci 13:317–323
12. Arrow KJ (2012) Social choice and individual values. Yale University Press
13. Funtowicz SO, Ravetz JR (1991) A new scientific methodology for global environmental issues. Ecol Econ Sci Manag Sustain 10:137
14. Renn O (2015) Stakeholder and public involvement in risk governance. Int J Disaster Risk Sci 6:8–20
15. Stirling A (1998) Risk at a turning point? J Risk Res 1:97–109

16. Stirling A (2006) Analysis, participation and power: justification and closure in participatory multi-criteria analysis. Land Use Policy 23:95–107
17. Banville C, Landry M, Martel J, Boulaire C (1998) A stakeholder approach to MCDA. Syst Res Behav Sci Off J Int Fed Syst Res 15:15–32
18. Petts J, Leach B (2000) Evaluating methods for public participation: literature review. Environ Agency Bristol
19. Stagl S (2007) Emerging methods for sustainability valuation and appraisal. Final report, SDRN-Sustainable Development Research Networks
20. Vari A (1995) Citizens' Advisory Committee as a model for public participation: a multiple-criteria evaluation. In: Fairness and competence in citizen participation. Springer, pp 103–115
21. Burgess J, Stirling A, Clark J et al (2007) Deliberative mapping: a novel analytic-deliberative methodology to support contested science-policy decisions. Public Underst Sci 16:299–322
22. Gregory R, Keeney RL (1994) Creating policy alternatives using stakeholder values. Manag Sci 40:1035–1048
23. McDowall W, Eames M (2007) Towards a sustainable hydrogen economy: a multi-criteria sustainability appraisal of competing hydrogen futures. Int J Hydrog Energy 32:4611–4626
24. Munda G (1996) Multicriteria analysis for land-use management
25. Proctor W, Drechsler M (2006) Deliberative multicriteria evaluation. Environ Plan C Gov Policy 24:169–190
26. Renn O, Webler T, Rakel H et al (1993) Public participation in decision making: a three-step procedure. Policy Sci 26:189–214
27. Stagl S (2006) Multicriteria evaluation and public participation: the case of UK energy policy. Land Use Policy 23:53–62
28. Stirling A, Mayer S (2001) A novel approach to the appraisal of technological risk: a multicriteria mapping study of a genetically modified crop. Environ Plan C Gov Policy 19:529–555
29. Hill M (1968) A goals-achievement matrix for evaluating alternative plans. J Am Inst Plann 34:19–29
30. Hill M (1973) Planning for multiple objectives. An approach to the evaluation of transportation plans
31. Barfod MB (2012) Optimising transport decision making using customised decision models and decision conferences. Kongens Lyngby, Denmark DTU Transport
32. Barfod MB, Salling KB (2015) A new composite decision support framework for strategic and sustainable transport appraisals. Transp Res part A Policy Pract 72:1–15
33. Jensen AV (2012) Appraisal of transport projects: assessing robustness in decision making
34. Jensen A V, Salling KB, Leleur S (2013) The Sustain Appraisal Framework: flexible decision support for national sustainable transport planning. In: 13th world conference of transport research', Rio de Janeiro, Brazil, July. pp 15–18
35. Leleur S (2012) Complex strategic choices: applying systemic planning for strategic decision making. Springer Science & Business Media
36. Salling KB, Jensen AV, Leleur S (2005) COSIMA-DSS evaluation system: a new decision support system for large-scale transport infrastructure projects. Publishing House of Poznan University of Technology, Pozn Poland, pp 601–607
37. Salling KB, Barfod MB, Pryn MR, Leleur S (2018) Flexible decision support for sustainable development: the SUSTAIN framework model. Eur J Transp Infrastruct Res 18
38. Macharis C, De Witte A, Ampe J (2009) The multi-actor, multi-criteria analysis methodology (MAMCA) for the evaluation of transport projects: theory and practice. J Adv Transp 43:183–202
39. Macharis C, De Witte A, Turcksin L (2010) The Multi-Actor Multi-Criteria Analysis (MAMCA) application in the Flemish long-term decision making process on mobility and logistics. Transp Policy 17:303–311. https://doi.org/10.1016/J.TRANPOL.2010.02.004
40. Macharis C, Nijkamp P (2011) Possible bias in multi-actor multi-criteria transportation evaluation: issues and solutions. Res Memo 31

41. Macharis C, Nijkamp P (2013) Multi-actor and multi-criteria analysis in evaluating mega-projects. In: International handbook on mega-projects. Edward Elgar Publishing
42. Dimitriou HT, Harman R, Ward EJ (2010) Incorporating principles of sustainable development within the design and delivery of major projects: an international study with particular reference to major infrastructure projects. OMEGA Centre, University College London, London, UK
43. Ward EJ, Dimitriou HT, Dean M (2016) Theory and background of multi-criteria analysis: toward a policy-led approach to mega transport infrastructure project appraisal. Res Transp Econ 58:21–45
44. De Brucker K, Macharis C, Verbeke A (2015) Two-stage multi-criteria analysis and the future of intelligent transport systems based safety innovation projects. IET Intell Transp Syst 9:842–850
45. Dean M (2018) Assessing the applicability of participatory multi-criteria analysis (MCA) methodologies to the appraisal of mega transport infrastructure
46. Dean M (2020) A practical guide to multi-criteria analysis. Adv Transp Policy Plan 6:165–224
47. Forester J (1999) The deliberative practitioner: encouraging participatory planning processes. Mit Press
48. Healey P (1998) Collaborative planning in a stakeholder society. Town Plan Rev 1–21
49. Healey P (2003) Collaborative planning in perspective. Plan Theory 2:101–123
50. Innes J (2018) Planning through consensus building: a new view of the comprehensive planning ideal. In: Classic readings in urban planning. Routledge, pp 147–161
51. Dean M (2020) Multi-criteria analysis. In: Advances in transport policy and planning. Elsevier, pp 165–224
52. Lami IM, Masala E, Pensa S (2011) Analytic network process (ANP) and visualization of spatial data: the use of dynamic maps in territorial transformation processes. Int J Anal Hierarchy Process (IJAHP) Creat Decis Found 3:92–106
53. Lami IM, Abastante F, Bottero M et al (2014) Integrating multicriteria evaluation and data visualization as a problem structuring approach to support territorial transformation projects. EURO J Decis Process 2:281–312. https://doi.org/10.1007/S40070-014-0033-X
54. Pensa S, Masala E, Lami IM (2013) Supporting planning processes by the use of dynamic visualisation. In: Planning support systems for sustainable urban development. Springer, pp 451–467

Chapter 5
Methodological Approach

Abstract This chapter outline a methodological approach to evaluating stakeholder preferences and decision-making dynamics in large-scale infrastructure projects. The proposed model integrates multi-criteria analysis (MCA) with a novel 30-year evaluation framework, enabling the comparative assessment of alternatives at different point in time. The MCA approach prioritises impact identification and stakeholder engagement through iterative roundtable discussions. By synthesising multiple evaluation techniques, the methodology facilitates consensus building, mediates competing interests, and accommodates temporal variability. This tailored framework enhances strategic decision-making, ensuring that sustainable infrastructure planning is informed by both technical assessments and political considerations.

Achieving a balance between stakeholder preferences and emerging alternatives requires the development of a model that can weigh up multiple legitimateinterest, which are often in competition with each other. Within this framework, evaluation serves as a technical framework, providing rational input to the decision-making process, while choice represents the synthesis of political and technical aspects.

However, in the initial strategic planning phases, Multi-Criteria Analysis (MCA) preference classification may be inadequate, as the alternative involved are often not precisely defined. Consequently, any resulting rankings would be based on imprecise preliminary information. This circumstance could lead to the unjustified rejection of potentially viable alternatives, causing decision-makers to become "locked in" to suboptimal project alternatives—a frequently encountered challenge in major transport projects [1, 2]. Therefore, evaluation at this initial stage serves to identify impacts that are crucial to stakeholders, not merely to rank alternatives. These findings can be used for more than just selecting or rejecting alternatives, facilitating the analysis of decision-maker dynamics and conflict resolution.

This evaluative context demands a bespoke MCA approach. The evaluation methodology proposed in this book is novelin that it synthesises multiple existing methods (several of which were presented in Chap. 3) into an original framework specifically tailored to the case study at hand.

E. M. Bertolini, *Temporal Value in MCDM*, Lecture Notes in Intelligent Transportation and Infrastructure, https://doi.org/10.1007/978-3-031-95485-6_5

The evaluation process involved direct and active engagement with all stakeholders through successive roundtable discussions designed to facilitate consensus building. This participatory approach is firmly rooted in both planning and multi-criteria analysis (MCA) evaluation traditions, and the advantages of integrating participatory methods with MCA analytical tools are recognised extensive literature (for a comprehensive review, see Stagl [3]).

This methodological framework has proven particularly suitable as it synthesises multiple techniques for aggregating individual stakeholder perceptions, mediating and reconciling diverse viewpoints, effectively representing power dynamics among stakeholders, and—crucially—accommodating a multi-temporal context where alternatives must be evaluated across short-, medium- and long-term horizons.

5.1 Development of the Evaluation Method

From a methodological perspective, the primary development was a bespoke evaluation model, specifically designed to address the demands of assessing large-scale infrastructure within a multi-stakeholder decision-making context, through the integration of multiple MCA approaches.

The procedure is grounded in the characterization of the decision-making context through comprehensive analysis of the evaluation object, stakeholder roles, and their proposed alternatives. The methodology unfolds across three distinct phases: The initial phase focuses on establishing evaluative criteria through the construction of an assessment matrix, leading to the classification of individual stakeholder preferences. The second phase involves projecting alternative performance across temporal dimensions. The third phase reconciles individual preference systems into a collective group preference framework through the analysis of strategic issues that emerged during working group sessions.

The transition from multiple individual datasets to a unified dataset is achieved through the harmonization and optimization of diverse factors spanning multiple domains. This creates a unified evaluative procedure that simultaneously addresses choice-related issues in group decision-making contexts, the temporal value of proposed alternatives, and considerations across economic, social, and environmental dimensions. It is noteworthy that the challenge of integrating such diverse values within a single study, while preserving their distinct identities, presents not only methodological challenges but also philosophical and conceptual complexities.

The framework for analyzing individual preferences is structured through the following sequential steps:

1. Ordering of alternatives based on individual preferences, incorporating identified criteria and their respective weightings for each decision-maker, For a more extensive and in-depth explanation, please refer to Chap. 7, while practical applications are in Chap. 11.

2. Analysis of individual rankings considering the temporal value of alternative performance, A detailed and comprehensive explanation can be found in Chap. 8, whereas Chap. 11 contains the practical applications.;
3. Balancing of results through group decision-making processes, see Chap. 9, for a thorough explanation, while Chap. 11 provides a concise summary of the application.

This framework represents a comprehensive approach to synthesizing diverse stakeholder perspectives while maintaining methodological rigor and temporal awareness in infrastructure evaluation.

The proposed methodological framework represents a significant advancement in infrastructure evaluation by addressing a critical gap in contemporary Multi-Criteria Analysis applications. While traditional MCA approaches often treat temporal aspects and stakeholder perspectives as separate domains, this methodology innovatively integrates them into a unified analytical framework. This integration is particularly noteworthy as it allows for the simultaneous consideration of short-term implementation challenges, medium-term operational impacts, and long-term strategic outcomes—a comprehensive temporal perspective that is often lacking in conventional evaluation approaches.

The approach's robustness is enhanced by its iterative nature, allowing for continuous refinement of preferences and criteria weights as stakeholders gain deeper insights into the project's implications across different time horizons. This iterative dimension is particularly valuable in complex infrastructure projects where initial stakeholder preferences might evolve as understanding of long-term impacts deepens. The methodology's structure accommodates such evolution while maintaining analytical consistency, thereby ensuring that the evaluation process remains both rigorous and responsive to emerging insights.

The framework's integration of temporal considerations represents a departure from traditional MCA applications, which often struggle to meaningfully incorporate time-dependent variables. By explicitly modeling how stakeholder preferences and project impacts evolve over time, the methodology provides a more nuanced and realistic basis for decision-making. This temporal integration is achieved without sacrificing the methodology's adaptability to different contexts, making it particularly valuable for diverse infrastructure projects ranging from transport systems to urban development initiatives.

References

1. Cantarelli CC, Flybjerg B, Molin EJE, van Wee B (2013) Cost overruns in large-scale transportation infrastructure projects: explanations and their theoretical embeddedness
2. te Boveldt G, Keseru I, Macharis C (2022) When monetarisation and ranking are not appropriate. A novel stakeholder-based appraisal method. Transp Res Part A Policy Pract 156:192–205. https://doi.org/10.1016/J.TRA.2021.12.004
3. Stagl S (2007) Emerging methods for sustainability valuation and appraisal. Final report, SDRN-Sustainable Development Research Networks

Chapter 6
The Working Group

Abstract Evaluating the temporal value of infrastructure projects requires a structured decision-making framework that considers the views of various stakeholder. This chapter examines the role of the Working Group as a central mechanism in the proposed evaluation model, which facilitates stakeholder engagement across multiple territorial levels. Comprising of institutional and managerial entities, the group ensures a balanced approach to decision-making by aligning public administration objectives with private sector interests. The model emphasizing multilevel governance and incorporates mechanisms for conflict resolution and preference aggregation, fostering a nuanced synthesis of stakeholder perspectives. This methodological approach enhances the transparency and effectivenessof decision-making, ensuring that infrastructure projects meet regulatory, economic, and social sustainability criteria.

The Working Group plays a pivotal role as an essential component of the proposed evaluation model. This entity is responsible for analyzing user needs and territorial constraints specific to the study object, gathering information, researching, evaluating, and selecting various proposed alternatives. Its primary objective is to formulate decisions through careful examination of diverse perspectives.

In this context, decision-makers (individuals formally charged with decision-making authority) comprise both Public Administration bodies and a joint-stock company, engaged in the decision-making process across different territorial levels (local, municipal, regional) with distinct objectives. Within the working group framework, the evaluator's role is to facilitate dialogue and manage relationships among decision-makers. The group's composition is therefore crucial for identifying optimal solutions (Fig. 6.1).

Please check and confirm if the inserted citation of 'Fig. 6.1' is correct. If not, please suggest an alternate citation. Please note that figure should be cited sequentially in the text.checked and suggested alternative citation

According to Li et al. stakeholders in major sustainable infrastructure projects can be defined as "those who can influence the project process and/or final outcomes, whose living environments are positively or negatively affected by the project, and

E. M. Bertolini, *Temporal Value in MCDM*, Lecture Notes in Intelligent Transportation and Infrastructure, https://doi.org/10.1007/978-3-031-95485-6_6

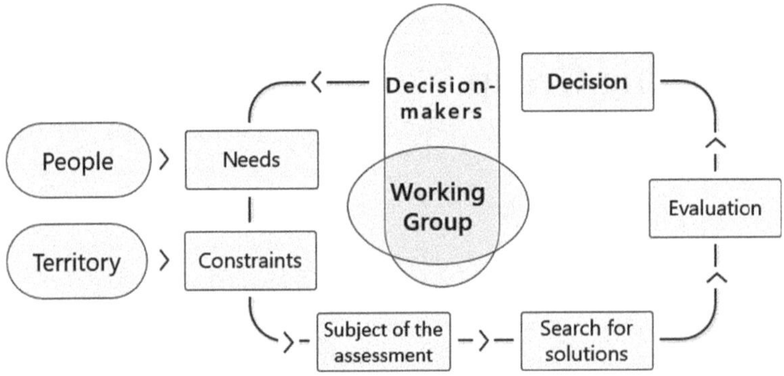

Fig. 6.1 Interaction between the working group, decision-making process, territorial dynamics, user needs, constraints, and evaluation. The diagram illustrates the interconnected relationships among stakeholders, highlighting how evaluators mediate between decision-makers, territorial demands, and regulatory constraints to ensure balanced and informed decision-making

who receive associated direct and indirect benefits and/or losses" [1]. To achieve the most realistic understanding of current dynamics, and in accordance with current transport and planning regulations.

The working group comprises key institutional and managerial entities involved in the governance and strategic development of the freight terminal and border infrastructure.

The management entity oversees the administration and operation of the freight village and border station, ensuring the efficiency of logistics activities and the integration of transport services within the broader territorial framework.

The local government authority is responsible for land-use planning and territorial management, acting as the owner of the real estate assets associated with the intermodal hub. It plays a fundamental role in aligning infrastructure development with municipal policies and community needs.

The port system authority is tasked with coordinating and managing strategic decisions related to port and intermodal logistics in compliance with national regulations. Its role is to facilitate the integration of maritime and inland transport networks, ensuring the coherence of regional and international trade flows.

The regional government exercises regulatory and supervisory functions in urban and territorial planning. It is responsible for defining strategic priorities, coordinating infrastructure investments, and monitoring the alignment of local projects with broader regional and national policies.

Together, these entities contribute to a multi-level governance framework, balancing operational management with policy-driven decision-making to enhance the efficiency and competitiveness of the logistics and transport system.

This stakeholder framework reflects the complex interplay of public and private interests in large-scale infrastructure development, highlighting the multi-level governance structure typical of contemporary infrastructure projects. The inclusion of diverse institutional actors at various territorial levels ensures comprehensive consideration of local, regional, and systemic impacts while acknowledging the distinct roles and responsibilities of each stakeholder in the decision-making process.

The deliberate structuring of stakeholder involvement through this working group model facilitates systematic evaluation of alternatives while ensuring that various institutional perspectives and regulatory requirements are adequately represented in the decision-making process. This approach aligns with contemporary best practices in infrastructure planning, where multi-stakeholder engagement is increasingly recognized as crucial for project success and sustainability.

The methodology's approach to stakeholder dynamics represents a sophisticated response to the complex power relationships inherent in large infrastructure projects. This consideration is particularly crucial in infrastructure planning, where the interplay between public authorities, private interests, and community stakeholders frequently creates complex dynamics that can significantly influence project outcomes.

The approach includes mechanisms for identifying and addressing potential conflicts early in the evaluation process, rather than allowing them to emerge as obstacles during later project phases. This proactive conflict management aspect is integrated into the preference aggregation process, where divergent stakeholder interests are not merely averaged or simplified but are carefully analyzed to understand underlying concerns and identify opportunities for meaningful compromise. This sophisticated treatment of stakeholder dynamics helps ensure that the final decisions reflect a genuine synthesis of stakeholder perspectives rather than merely representing the preferences of the most influential actors.

Reference

1. Li THY, Ng ST, Skitmore M (2012) Conflict or consensus: an investigation of stakeholder concerns during the participation process of major infrastructure and construction projects in Hong Kong. Habitat Int 36:333–342

Chapter 7
Ranking of Alternatives Based on Individual Preferences

Abstract This phase enables the ranking of alternatives according to each stakeholder's individual preferences, as expressed through criteria and weights. Since each stakeholder has different objectives, they are asked to express these through specific criteria in order to identify the alternative that can maximize these objectives from different perspectives (economic, technical, social, environmental). First, the chapter will analyse the role of criteria, then propose the technique for determining weights: simplified Simos method.

In strategic decision-making processes involving multiple stakeholders, the search for equilibrium between individual preferences and available alternatives necessitates a robust methodological framework. This chapter examines the first phase of the proposed evaluation model—the ranking of alternatives according to individual preferences. Unlike traditional approaches that focus solely on producing a final ranking, this methodology recognizes that in early strategic planning stages, project alternatives are often not well-defined. Consequently, rankings based on preliminary information may unjustifiably eliminate potentially valid options, leading to suboptimal decisions—a common issue in large transport infrastructure projects.

The approach presented here aims not merely to classify alternatives but to identify impacts deemed crucial by stakeholders and to analyze decision-maker dynamics, potentially resolving conflicts through a participatory process. This requires constructing a model that balances multiple legitimate, sometimes competing interests, where evaluation provides technical rationality while choice represents the synthesis of political and technical aspects.

This evaluation model combines multiple MCA approaches, structured into three main phases: (1) defining evaluation criteria and constructing evaluation matrices for individual preferences; (2) projecting alternative performances over time; and (3) reconciling individual preference systems into a group preference system through strategic issue analysis.

The transition from multiple individual datasets to a single cohesive system occurs through the harmonization and optimization of diverse factors spanning economic, social, and environmental domains. This creates a unified evaluative procedure

E. M. Bertolini, *Temporal Value in MCDM*, Lecture Notes in Intelligent Transportation and Infrastructure, https://doi.org/10.1007/978-3-031-95485-6_7

that addresses group decision-making challenges while considering the temporal value of alternatives. The difficulty of associating such diverse values within a single study, without compromising their identity, presents both methodological and philosophical-conceptual challenges.

Among multi-attribute methods, which analyze a finite number of predetermined alternatives through a deterministic approach, linear additive models have been selected as the foundational analytical framework. These models combine values of an option across multiple criteria by multiplying each criterion score by its respective weight and summing all weighted scores. This approach is appropriate when preference criteria are independent of each other and uncertainty is not formally incorporated into the MCA model. Most MCA approaches utilize this additive model, which forms the basis of Multiple Criteria Decision Method (MCDM).

Linear additive models have demonstrated effectiveness in providing robust support to decision-makers across various problems and circumstances. The Simple Additive Weighing method (SAW), represents one of the most intuitive and widely used approaches for addressing multi-criteria decision problems. Its mathematical formulation identifies the preferred alternative (A*) through the following equation:

$$V_i(a) = \sum_{i=1}^{n} w_r p_{ir}(a_i) \tag{7.1}$$

where:

$V_i\ (a)$ is the total value of an alternative;

a_i is the ith alternative;

$p_{ir}\ (a_i)$ reflects the performance of an alternative according to criterion r;

w_i is the weight assigned to reflect the importance of criterion r.

This approach assumes all criteria are independent, making it particularly suitable for the initial phase of the evaluation model proposed in this research.

7.1 Criteria

The ranking of alternatives based on individual preferences requires establishing criteria and determining their weights for each decision-maker. Since each stakeholder has different objectives, they must express these through specific criteria to identify the alternative that maximizes their goals across different perspectives (economic, technical, social, environmental).

As stated in Chap. 2, in this context, a judgment criterion is defined as the element to be determined in order to pursue a given objective. Since each stakeholder has different objectives, they are asked to express them through these criteria to identify, from different perspectives (economic, technical, social, environmental), the alternative that maximizes them.

Criteria, which are essential in the evaluation phase, serve a dual purpose: on the one hand, they produce easily measurable parameters that provide essential information for decision-makers; on the other hand, they help manage the complexity of the alternatives.

The identification of these criteria was carried out through the direct involvement of stakeholders who articulate their objectives and priorities, moreover these criteria span multiple dimensions (economic, technical, social, environmental) to ensure comprehensive evaluation. These are then synthesized into qualitative judgment criteria, which are useful for evaluating alternatives based on relevant aspects and characteristics. These criteria, being qualitative and case-dependent, are grouped by stakeholder, as they directly reflect their policies and concerns.

The principle of compensation among criteria was deemed unacceptable for the following reasons:

- A transport system intervention often generates impacts across multiple domains (beyond transportation itself), ranging from socioeconomic to environmental and landscape-related aspects. The plurality of objectives set with reference to different areas of interest—sometimes even conflicting—renders the criteria unsuitable for compensation [1];
- Moreover, Multi-Criteria Analysis (MCA) methods, which rely on a systematic comparison of different alternatives, emphasize the non-compensability of criteria [2].

The numerical values assigned to qualitative criteria are based on the judgment of an external expert involved in the evaluation: Prof. Giovanni Longo, Full Professor of Transport at the University of Trieste, specializing in mobility issues, transport systems, and logistics.

These values are assigned considering the performance that each alternative might exhibit for each criterion. A rating scale from 0 to 5 was established based on the significance of the effects of the *i*th alternative on the *j*th criterion. These values, therefore, represent the extent to which each alternative has significant effects on the territory according to the following scale (Table 7.1).

Mathematically, the set of alternatives is defined as A: = {Aa} a ∈ {1, …, 5} and the set of criteria as C: = {Cj} j ∈ {1, …, n}}. ach criterion Cj is a function mapping the set of alternatives *A* to the set {0, …, 5} i.e., Cj: A → {0, …,5}.

Table 7.1 Performance value scale

Performance value scale	
0	No or negligible effects
1	Weak positive effect
2	Moderate positive effect
3	Strong positive effect or local-scale impact
4	Very strong positive effect or medium-scale impact
5	Significant positive effect or large-scale impact

We define P_{aj} as the value Cj (Aa); this value represents the performance of alternative *Aa* for criterion Cj. Through this approach, each alternative Aa can be classified based on its value P_{aj} for each stakeholder.

7.2 Technique for Weight Determination: Simplified Simos Method

A significant factor in multi-criteria decision support models (MCA) is the weighting of criteria, or the importance assigned to criteria. The weight assigned to a criterion serves as a scaling factor that enables comparison between criteria in determining the decision maker's preference state [3]. Generally, these parameters incorporate the preferences of a single decision maker (DM) into the model.

In evaluating the object of study, after careful examination of the MCA methods employed here, and to overcome the identified criticisms, we selected a methodology called the "Revised Simos' Procedure" proposed by Figueira and Roy [4], but implemented through a simplified calculation procedure proposed by Defrancesco [3].

7.2.1 State of Art

Existing methods widely used to evaluate the importance weights of criteria can be classified into two categories: (i) direct assessment procedures, in which the DM is asked to explicitly express criteria weights in percentage terms, and (ii) indirect assessment procedures, which deduce weights from pairwise comparisons of criteria or reference alternatives [5].

The second category of methods includes, among others:

- The cards method proposed by Simos that will be described in the following section [6, 7];
- The centralized weights method, which requires the DM to make a series of ordinal criteria comparisons that are formulated as linear inequalities [8];
- The TACTIC method, in which the relative importance of criteria is evaluated as a system of functional representations of relationships [9];
- DIVAPIME, which is implemented by making pairwise comparisons of fictitious alternatives and has been adapted from ELECTRE methods [10];
- The Analytic Hierarchy Process (AHP), proposed by Saaty, in which the DM is asked to provide pairwise comparisons on the priority of criteria on a predetermined numerical scale [11];
- MACBETH, which deduces weights as attractiveness values from pairwise comparisons of criteria on a qualitative scale, thus measuring the magnitude of attractiveness [12].

- Bisdorff, Meyer, and Veneziano proposed a mixed integer linear programming model to deduce criteria importance weights from overall outranking statements, maximizing the stability of the outranking graph [13].

7.2.2 *Simos Method*

The method proposed by Jean Simos in 1990 has gained popularity and has been applied to various types of problems due to its simplicity and the convenience it offers decision makers (DMs) in expressing their preferences. Specifically, it requires constructing a hierarchy of evaluation criteria by involving the DM in a "playing card" procedure to assign numerical values to criteria.

The original Simos method consists of three steps concerning interaction with the DM and information gathering:

1. The DM receives a deck of cards with the name of a criterion on each (n cards, each corresponding to a specific criterion of a family F) along with some blank cards.
2. The DM is asked to rank the cards/criteria from least important to most important, arranging them in ascending order. If multiple criteria have the same importance, the DM should create a subset by keeping the corresponding cards together with a clip.
3. Finally, the DM is asked to insert blank cards between two successive cards (or subsets of equally important criteria) if they believe the difference between them is wider. The greater the difference between the weights of criteria (or subsets of criteria), the greater the number of blank cards that should be placed between them. Specifically, if u denotes the difference in value between two successive criterion cards, then one blank card signifies a difference of twice u, two blank cards signify a difference of three times u, and so forth.

The information provided by the DM is used by the Simos method to determine weights according to the following algorithm:

1. Ranking of subsets of equally important criteria from least to most important, also considering blank cards,
2. Assigning a position to each criterion/card and to each blank card,
3. Calculating non-normalized weights,
4. Determining normalized weights.

The least qualified card is assigned Position 1, while the most qualified is assigned Position n. The non-normalized weight of each rank/subset is determined by dividing the sum of positions of a rank by the total number of criteria belonging to it. The non-normalized weights are then divided by the total sum of the positions of criteria in each rank (excluding blank cards); to normalize them, the values thus obtained are rounded to the nearest or higher integer value.

However, several criticisms have emerged: both Scharling [14] and Figueira and Roy [4] expressed objections to how the Simos procedure determines weights. One of the main problems reported is that it elicits only one set of weights that satisfies the model expressed by the DM. However, other sets of weights could potentially better fit the DM's preferences regarding the relative importance of criteria. Such sets of weights cannot be obtained with the Simos procedure.

A second critical point is that the procedure processes criteria with the same importance (i.e., the same weight) in a non-robust manner. If one attempts to reorder the cards between two subsets, the distance (difference in weights) between successive subsets changes in an uncontrolled way. This phenomenon occurs because the weight difference between two successive subsets of criteria is automatically influenced by the number of cards existing in these subsets. The user, however, has no real or absolute perception of how numerical values are determined by the procedure. Finally, Figueira and Roy do not support rounding normalized weights to 100, as they perceive it as an unrealistic process.

In their effort to address these issues, Figueira and Roy [4] proposed a revised version of the Simos method. In addition to the three-phase data collection process, the new procedure introduces a fourth phase, which requires the DM to indicate how many times the last criterion is more important than the first in the ranking. This ratio is defined as the z-ratio. It is used to define a fixed interval between the weights of criteria or their subsets. The variable u denotes this interval: $u = (z - 1)/e$, where e is the number of different weight classes (i.e., single card, subsets of cards, and blank card).

The Simos method, although showing considerably easy, almost naive, data collection and implementation, has been widely used in scientific literature. Several authors have employed the method, mostly combined with ELECTRE-type methods, to evaluate the importance of criteria weights. It is known that Figueira and Roy initiated a broad application of the Simos methodology, as most of the aforementioned studies make use of their revised version.

7.2.2.1 Some Examples of Simos Method Application in Infrastructure

Mohamed Marzouk, Ahmed Nouh, and Moheeb El-Said used the Simos procedure to construct an evaluation system on how infrastructure projects—particularly green bridge projects—can significantly contribute to sustainable development. Since there is little existing work on appropriate methods to evaluate the sustainability performance of projects, criteria weights are defined based on various building industry standards to decide which criteria influence the sustainability of bridge projects. The initial list of criteria is identified from unstructured interviews; subsequently, structured interviews and survey questionnaires are conducted to identify the final list deemed important for judging their status with respect to sustainability. After acquiring the final list of important criteria, which are: life cycle cost analysis criterion, sustainable site selection, intelligent transportation systems, storage/separation areas, and regional materials, respectively, they proceeded to the second step. The second step consists of acquiring the normalized weights between each category as presented. Finally, the global weights of criteria are obtained by multiplying the local weights of criteria by the respective category weight. After obtaining the global weights with the Simos procedure, a rating system for green bridges is proposed [15].

José Roberto Ribas, Giuliana C. R. Severo, Leticia C. G. Felix, and Kim P. C. Perpetuo, regarding dam safety-focused procedures, primarily due to rules established by various regulatory bodies, used the Simos method combined with a procedure adopted to convert scores into the paired comparison format. The MCA model was developed based on regulations and guidelines recommended for engineering works and common sense adopted in dam management and maintenance, in order to evaluate the major socioeconomic and environmental impacts that any plant incident can cause. The method allowed for the classification of safety procedures according to their priorities, thus providing plant management with elements to better program monitoring and staff training [16].

Eleftherios Siskos and Peter Burgherr, with their research work "Multicriteria Decision Support for Electricity Supply Resilience Assessment: Exploring Interacting Criteria," aim to support energy policy makers in Europe and provide guidelines and areas for improvement at the national level. The increasing risks of prolonged power supply interruptions and/or strong price fluctuations are indeed emphasizing the need for an assessment of electricity supply resilience. For this purpose, a multicriteria methodology is developed to evaluate electricity supply resilience, a complex and multidimensional concept that is receiving increasing attention at the European level. In particular, this overall assessment is established based on a synergy of two complementary multicriteria decision aid (MCA) methods. A synergy of the Choquet integral and an updated cards method is applied; a procedure based on the Simos method supports the robust elicitation of Choquet capacities, thus canceling their arbitrary effects on the final benchmarking [17].

7.2.3 *Simplified Simos Method*

The simplified calculation procedure proposed by Crisman et al. [18] and utilized for the present evaluation model unfolds according to the following scheme, with the procedure explained step by step below:

1. Criteria are ordered from least important to most important with the possibility of establishing tied rankings. The notation r (r = 1, ..., r) indicates the number of positions (steps) in the ordering, and c_r to indicate the number of parameters present in each step.
2. One or more "blank cards" are inserted into the ordering defined in point 1 to represent differences in importance between parameters placed in contiguous positions. The insertion of a "blank card" signifies that—for the decision maker—the difference in importance between two parameters placed in contiguous positions in the ordering is twice the difference that would exist in the absence of a "blank card"; if two "blank cards" are inserted, the importance is triple, and so on.

3. The value one is assigned to the first step containing the least important criterion according to the ordering, two to the second, and so on up to the step with the most important criterion/criteria; steps occupied by "blank cards" are also considered in the numbering.
4. The cells containing "blank cards" are removed from the numbered ordering; in this way, the values associated with each step vr represent an initial, non-normalized judgment of the relative importance of the various parameters considered.
5. An importance ratio (z) is defined between the most important criterion and the least important one.
6. The weight wrz is calculated through linear interpolation in such a way as to obtain a ratio between the weight associated with the most important parameter and the least important one equal to z:

$$w_r^z = \frac{v_r(z-1)}{v_{rmax} - v_{rmin}} + \frac{v_{rmax} - zv_{rmin}}{v_{rmax} - v_{rmin}} \tag{7.2}$$

where:

w_r^z is the value of the criterion weight;

v_r is an initial judgment not normalized with respect to the importance of the criterion;

7. For each step r of the ordering, the product between cr and wrz is calculated, and the products obtained are summed:

$$w_t^z = \sum_r w_r^z c_r \tag{7.3}$$

8. The weight w_r^n is calculated by normalizing w_r^z by w_t^z for each criterion belonging to each step r of the ordering:

$$w_r^n = \frac{w_t^z}{w_r^z} \tag{7.4}$$

where:

w_r^n is the normalized value of the criterion weight;

w_t^z is the value of the step weight;

w_r^z is the value of the criterion weight.

It is noted that the method used presents a series of advantages compared to Simos' original method, as the ordering chosen by the decision maker does not present robustness problems; the introduction of "blank cards" allows for the insertion of different weights for criteria independently of their number.

Similarly, the parameter z allows for the assignment of weights independently of any ties; the simplification of the calculation proposed by Defrancesco et al. [3] makes the procedure intuitive and easy to implement with a standard spreadsheet.

7.3 Conclusion

This approach recognizes that in strategic decision-making contexts, particularly in early planning phases, the primary objective is not merely to rank alternatives but to understand stakeholder priorities and identify crucial impacts. The results serve not only for alternative selection but also for analyzing stakeholder dynamics and resolving potential conflicts through a participatory, deliberative process.

This systematic process enables the identification of each stakeholder's preferred alternatives based on their specific objectives and criteria weights. The resulting individual rankings form the foundation for subsequent phases of the evaluation model, where individual preferences are reconciled into a group preference system through strategic issue analysis.

The ranking of alternatives based on individual preferences constitutes the first phase of a comprehensive evaluation model designed to address complex decision-making scenarios involving multiple stakeholders and competing objectives. By utilizing a structured approach to criteria definition, weight elicitation, and preference aggregation through linear additive models, this methodology enables the identification of each stakeholder's preferred alternatives based on their specific objectives.

This individual preference ranking phase does not aim to produce a definitive ranking but rather to illuminate stakeholder priorities and identify crucial impacts. The results provide a foundation for subsequent phases of the evaluation process, where individual preferences are reconciled into a group preference through temporal projection and strategic issue analysis system.

The methodology acknowledges that in strategic planning contexts, particularly in early phases, the primary objective is to understand stakeholder dynamics and identify potential conflicts rather than prematurely committing to specific alternatives. By combining technical rationality with participatory processes, this approach enhances the quality and legitimacy of strategic decisions while reducing the risk of "lock-in" to suboptimal alternatives—a common issue in large infrastructure projects.

References

1. Cartenì A (2016) Processi decisionali e Pianificazione dei trasporti. Lulu.com
2. Peng H, Xiao Z, Wang J, Li J (2021) A decision support framework for new energy selection in rural areas from the perspectives of information reliability and criterion non-compensation. Energy 235:121435. https://doi.org/10.1016/J.ENERGY.2021.121435
3. Defrancesco E, Rosato P, Rossetto L, et al (2006) Il risarcimento per danno ambientale: aspetti teorici e operativi della valutazione economica
4. Figueira J, Roy B (2002) Determining the weights of criteria in the ELECTRE type methods with a revised Simos' procedure. Eur J Oper Res 139:317–326. https://doi.org/10.1016/S0377-2217(01)00370-8
5. Siskos E, Tsotsolas N (2015) Elicitation of criteria importance weights through the Simos method: a robustness concern. Eur J Oper Res 246:543–553. https://doi.org/10.1016/j.ejor.2015.04.037

6. Simos J (1990) Levaluation environnementale: un proces sur cognitif negocie
7. Simos J (1990) Evaluer l'impact sur l'environnement: une approche originale par l'analyse multicritère et la négociation. Presses polytechniques et universitaires romandes, Lausanne
8. Solymosi T, Dombi J (1986) A method for determining the weights of criteria: the centralized weights. Eur J Oper Res 26:35–41
9. Vansnick J-C (1986) On the problem of weights in multiple criteria decision making (the noncompensatory approach). Eur J Oper Res 24:288–294
10. Mousseau V (1995) Eliciting information concerning the relative importance of criteria. In: Advances in multicriteria analysis. Springer, pp 17–43
11. Saaty TL (1996) Decision making with dependence and feedback: the analytic network process. RWS Publications Pittsburgh
12. Bana e Costa CA, de Corte J, Vansnick J (2010) Macbeth (measuring attractiveness by a categorical based evaluation technique). Wiley Encycl Oper Res Manag Sci
13. Bisdorff R, Meyer P, Veneziano T (2014) Elicitation of criteria weights maximising the stability of pairwise outranking statements. J Multi-Criteria Decis Anal 21:113–124
14. Schärlig A (1996) Pratiquer Electre et Prométhée: un complément à" Décider sur plusieurs critères". PPUR presses polytechniques
15. Marzouk M, Nouh A, El-Said M (2014) Developing green bridge rating system using Simos' procedure. HBRC J 10:176–182
16. Ribas JR, Severo JCR, Felix LCG, Perpetuo KPC (2021) Prioritization of hydroelectric power plant earth dam safety procedures: a multi-criteria approach. Environ Earth Sci 80:1–13
17. Siskos E, Burgherr P (2022) Multicriteria decision support for the evaluation of electricity supply resilience: exploration of interacting criteria. Eur J Oper Res 298:611–626
18. Crisman B, Martellos L, Rosato P, Carlo M (2018) La valutazione degli investimenti per la sicurezza delle traverse urbane: un approccio gerarchico multiattributo, pp 1–16

Chapter 8
Evaluation of Temporal Value

Abstract Evaluating temporal value is critical for decision-making, especially in public policy contexts where resource allocation requires balancing short-, medium-, and long-term considerations. This chapter presents a theoretical framework for evaluating the temporal aspects of different alternatives, emphasising the importance of a structured approach to time-sensitive decision-making. It explores key methodologies for determining temporal value, such as social time preference and evaluation matrices, and critically reviews the current state of the art in this area. By addressing the complexities of intertemporal comparisons, this analysis lays the groundwork for more informed and sustainable policy decisions.

Public decision-making processes are becoming increasingly complex in a context where decisions must simultaneously account for economic, social, and environmental considerations. In this framework, certain countries, notably France, have witnessed a shift away from Multi-Criteria Decision Analysis (MCDA) towards single-criterion methods [1]. This transition has been primarily driven by significant discrepancies in key parameters, such as the value of time, which have led to inefficient allocation of government resources.

These challenges underscore the need for careful consideration when applying MCDA in policy-making contexts, particularly where resource allocation decisions have far-reaching economic implications. In this framework, decisions need to be evaluated across short, medium, and long-term horizons, as their planning timescales typically span several years or even decades. This multi-temporal approach is crucial, despite the fact that sustainable development aims to strike a balance between short-term and long-term action assessments.

The intricate nature of these decisions reflects the multifaceted challenges faced by modern policymakers. They must navigate a landscape where immediate needs often compete with long-term sustainability goals, necessitating a nuanced approach to decision-making. This complexity underscores the importance of developing robust methodologies that can effectively integrate diverse criteria across various time scales, thereby supporting more informed and sustainable public policy decisions [2].

E. M. Bertolini, *Temporal Value in MCDM*, Lecture Notes in Intelligent Transportation and Infrastructure, https://doi.org/10.1007/978-3-031-95485-6_8

The approach to the temporal dimension adopted for this research is as follows: alternatives are evaluated based on their performance in the short, medium, and long term. Furthermore, consideration is given to the duration over which the long-term effects may persist.

Indeed, a crucial issue in relation to transport infrastructure assessments is the analysis of aspects across sufficiently broad spatial and temporal scales [3]. Other fundamental aspects include the appropriate consideration of cumulative effects and meaningful monitoring [4].

8.1 Timeframe

This multi-temporal evaluation framework allows for a more comprehensive understanding of the potential impacts of transport infrastructure projects. By examining performance across different time horizons and considering the longevity of long-term effects, this approach provides a more nuanced perspective on the sustainability and efficacy of proposed alternatives. The emphasis on broad spatial and temporal scales, coupled with the consideration of cumulative effects and robust monitoring, aligns with best practices in infrastructure assessment. This holistic methodology enables decision-makers to better anticipate and manage the complex, long-term implications of transport infrastructure investments.

To account for effects that must encompass secondary, cumulative, and synergistic impacts over short, medium, and long terms—as indicated in the European Commission's Report to the European Parliament and Council on the application and effectiveness of the EIA Directive (Directive 85/337/EEC, as amended by Directive 97/11/EC)—the performance value of an interval is offset relative to the temporal interval itself.

Specifically, the short-term performance begins at the end of its designated period and extends until the medium-term performance reaches completion. The medium-term performance, in turn, persists until the long-term performance has fully established itself. Moreover, this long-term performance will continue to have enduring effects beyond its initial establishment.

This staggered approach to temporal performance evaluation allows for a more nuanced and comprehensive assessment of infrastructure impacts. By recognizing that the effects of interventions often manifest and evolve beyond their immediate timeframes, this methodology captures the complex, overlapping nature of infrastructure outcomes. It acknowledges that short-term actions may have delayed impacts, medium-term effects may persist into long-term horizons, and long-term consequences may extend far beyond the initial planning period.

Such a sophisticated temporal framework enables decision-makers to better anticipate and manage the cascading and cumulative impacts of infrastructure projects, aligning with the EU's emphasis on thorough environmental impact assessments. This approach supports more informed decision-making by providing a fuller picture

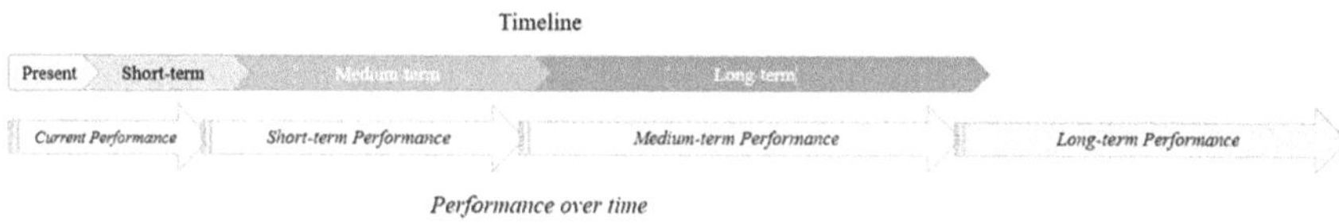

Fig. 8.1 Staggered approach to the temporal performance evaluation

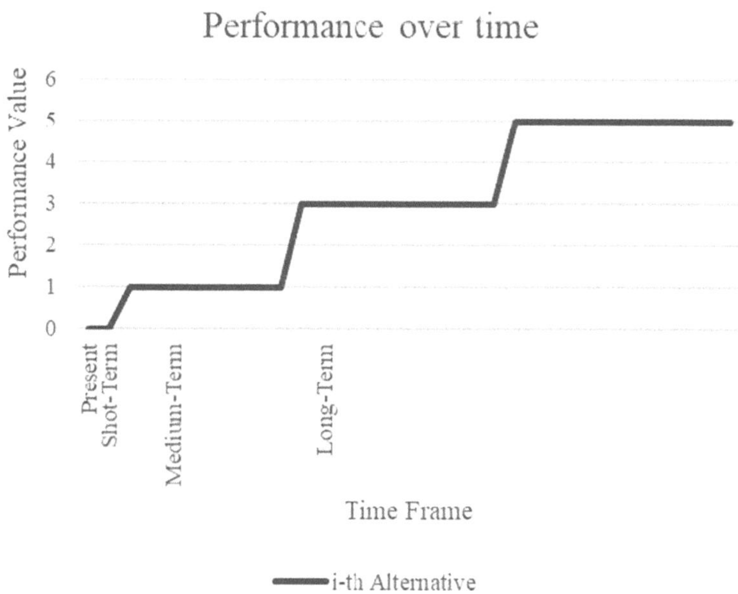

Fig. 8.2 Fictitious trend profile for the performance of the *i*th alternative

of how infrastructure performance unfolds and interacts across different time scales (Fig. 8.1).

The graph below presents a generic case where the alternative A_i demonstrates a trend of increasing performance (Fig. 8.2).

8.2 A Technique for Determining the Temporal Value

Like many public sector decisions, the present evaluation concerns choices to be made in the current moment that will have impacts manifesting in subsequent years. However, when comparing the performance of various alternatives that occur at different times and for different durations, significant difficulties arise from the need to aggregate all effects into a single measure.

Financial mathematics, however, offers a reasonably well-established aggregation procedure: discounting. This financial process allows for the determination of the present value of capital that has a future date as its natural maturity [5].

In this discipline, the financial value of time is associated with that of interest, or the reward for "foregoing" immediate consumption. Over time, through individual or business activities, monetary inflows or outflows occur, constituting cash flows that represent the amount of money expected to be paid or received over time as a consequence of financing or investment. These flows manifest with various maturities, and their values cannot be directly compared: it is therefore necessary to bring them to uniformity. The process by which future flows are converted to present flows allows for the determination of present value [6].

Discounting is thus essentially linked to the opportunity cost of capital: a given sum collected in two years is worth less than at the present time, because having that sum immediately represents a better economic-financial condition, as it can potentially be used to generate new useful capital through various possible forms of investment, or simply to spend. Therefore, it is necessary to scale down, or discount, the estimated future capital or each cash flow by an appropriate discount factor. This factor is calculated based on the discount rate and is greater the further the cash flow to be discounted is from the present time. In calculating the present value of capital, the formula is as follows:

$$NPV = \sum \frac{CF_j}{(1+r)^t} \quad (8.1)$$

where:

- NPV is the Net Present Value;
- CF_j are the cash flows at time j;
- r is the discount rate;
- t is the time of the cash flow [6].

This approach to temporal valuation enables a more coherent comparison of alternatives with differing time horizons and impact durations. By applying financial mathematics principles to public sector decision-making, it becomes possible to evaluate and compare long-term projects on a common basis, facilitating more informed and economically sound policy choices.

In Multi-Criteria Analysis (MCA), there is no equivalent technique to financial discounting, although in principle it is possible to adopt conventional discounting of monetary values and adapt it to physical impact indices other than monetary value. However, this is generally not done, more for cultural than epistemological reasons [7].

Nevertheless, time must be included in the evaluation to assess impacts that will occur over the years. The long-term vision, however, poses challenges for the decision-maker, as the performance of alternatives is uncertain, often difficult to measure, and the acceptability of decisions is harder to achieve.

Although MCA methods allow for the simultaneous consideration of potentially conflicting criteria, the state of the art remains limited regarding the balance between short-term and long-term action assessments. Only a few research studies offer frameworks for temporal aggregation [2].

Certainly, good decision facilitation practice would ensure that participants in any decision-making exercise have greater awareness of time-differentiated impacts through increased attention to their coherent arrangement in the evaluation.

This lack of a standardized approach to temporal considerations in MCA highlights a significant area for methodological development. While financial discounting provides a clear framework for comparing monetary values across time, the challenge lies in developing analogous methods for non-monetary criteria that are often central to public policy decisions. The need for such methods is particularly acute given the increasing emphasis on long-term sustainability and the complex, multi-faceted nature of many public policy challenges.

Addressing this methodological gap could significantly enhance the robustness of MCA in supporting long-term decision-making. It would enable decision-makers to more effectively balance short-term imperatives with long-term goals, thereby improving the overall quality and sustainability of policy outcomes. Future research in this area could focus on developing and validating temporal aggregation frameworks that can be consistently applied across various types of criteria, thus advancing the practical utility of MCA in complex, long-term decision contexts.

8.2.1 State of Art

Despite the importance of temporal (multi-period) evaluation of actions for sustainable decisions, only a few studies have addressed this aspect. Some authors consider long-term effects as a separate criterion [8, 9], while others employ scenario planning, predictive techniques, or fuzzy models to address future uncertainties [10].

As time must be incorporated into the definition of many other criteria to distinguish temporary from permanent consequences, one viable approach could be to model the time horizon as a separate criterion. This is typically done by explicitly stating the time horizon over which consequences are evaluated, with these horizons potentially differing from one criterion to another. An alternative possibility would be to conduct an MCA within an MCA, using expert judgments to assess the weights to be applied to impacts occurring in different future time [7].

In other studies [8, 11], long-term effects are evaluated approximately and/or qualitatively. Only recently have some temporal extensions of MCA methods been developed [9]. For example, in a forest management context, long-term impacts were considered as a specific criterion [9], with the local community asked to evaluate it. Khalili and Duecker propose a Sustainable Environmental Management System (SEMS) where actions are classified using ELECTRE III. In their study, they claim to have paid particular attention to criteria evaluation and carefully considered expected short-term and long-term consequences, though they omit explaining their method for reaching the obtained result.

Rout and Walshe propose a method that incorporates temporal preference information into multi-criteria decision analyses through criteria weights, ensuring a faithful representation of the decision-maker's preferences [12].

Frini and Benamor propose a multi-period multi-criteria method based on adapting TOPSIS to the temporal context, although the compensation between decision criteria on which TOPSIS is based (like scoring methods) is not appropriate for sustainability [13]. Meanwhile, Banamar and Smet generalise PROMETHEE to the temporal setting, applying weighted average through the aggregation of net flow scores across periods, thus using a compensatory method [14].

The literature review presented here shows a limited state of the art and a research area still largely to be developed regarding multi-period aggregation. As previously highlighted, compensation emerges as the main issue behind the few existing temporal proposals [2], with an ongoing debate between environmentalist and economist perspectives. However, it should not be overlooked that the most common method for ranking different temporal flows is to discount values according to the standard discount rate procedure [15].

This overview highlights the evolving nature of temporal considerations in Multi-Criteria Analysis. The incorporation of time as a distinct criterion or the use of nested MCA structures represents innovative approaches to addressing the complexities of long-term decision-making. However, the lack of standardised methodologies for integrating temporal aspects into MCA frameworks remains a significant challenge.

Future research in this area could focus on developing more robust and standardised methods for incorporating temporal considerations into MCA. This could include refining techniques for weighting future impacts, exploring the potential of fuzzy logic in long-term evaluations, and developing frameworks that can effectively balance short-term and long-term consequences in decision-making processes. Such advancements would significantly enhance the capability of MCA to support sustainable and forward-looking policy decisions.

This comprehensive review highlights the emerging field of temporal considerations in Multi-Criteria Analysis, emphasising the need for more robust methodologies to address long-term impacts in decision-making processes. The various approaches presented, from treating time as a separate criterion to developing temporal extensions of existing MCA methods, demonstrate the complexity of incorporating time into multi-criteria decision frameworks. The ongoing debate between compensatory and non-compensatory methods, particularly in the context of sustainability assessments, underscores the challenges in balancing short-term and long-term impacts. Future research in this area could focus on developing more sophisticated temporal aggregation techniques that can effectively capture the nuances of long-term consequences while maintaining the practical applicability of MCA in real-world decision-making scenarios.

8.2.2 *Social Time Performance*

We have thus far introduced the concept of the discount rate. It is therefore useful to examine how time can be treated in decision analyses through the choice of an appropriate discount rate; hence, the problem related to discounting is that of defining a social discount rate.

A social discount rate is understood as a discount rate that integrates:

- Economic-ethical considerations, such as the intergenerational, long-term nature (even over time spans of 50 years or more) of environmental costs and benefits, which is greater compared to market investments (typically with returns within 20–25 years);
- The "social" character of investment interventions, which may be linked to opportunity costs lower than those of the market. (There is a debate on this aspect with deep roots in economic theory [16]).

In all economic decision-making models (Cost–Benefit Analysis), the future is less important than the present; indeed, applying a discount rate attributes less importance to future impacts, whereas the environmentalist approach would tend to give greater importance to long-term impacts. The formal elicitation of time preference is laborious [15], thus assumptions are often made to approximate social time preference.

For individuals, time preference can be measured by the real interest rate on money lent or borrowed. People generally prefer to invest at fixed, low-risk rates, hoping to receive more than invested in the future to compensate for the deferral of consumption. These real rates of return give some indication of their individual time preference rate. Similarly, society as a whole prefers to receive goods and services sooner rather than later, known as social time preference.

In cost–benefit analysis, this is implemented by applying a discount rate to all future costs and benefits. For public projects, the discount rate used by the UK Treasury's appraisal guidance is known as the "Social Time Preference Rate" (STPR) and represents the value by which society values the present relative to the future [17].

The public sector discount rate conforms to social time preference, defined as the value that society attributes to present consumption relative to future consumption; it is based on utility comparisons between different points in time or different generations [17].

The Social Time Preference Rate (STPR) has two components:

1. "Time preference": the rate at which consumption and public spending are discounted over time, assuming per capita consumption does not change. This captures the preference for value now rather than later.
2. "Wealth effect": reflects the expected growth of per capita consumption over time, where future consumption will be higher than current consumption and is expected to have lower utility.

The Social Time Preference Rate is therefore expressed as:

$$r = \rho + \mu g \tag{8.2}$$

where:

- r represents the STPR
- ρ denotes the time preference, which encompasses both the pure time preference (δ) and catastrophic risk (L)
- μg signifies the wealth effect, calculated as the product of the marginal utility of consumption (μ) and the expected growth rate of future real per capita consumption (g).

This formulation provides a comprehensive framework for understanding the components that influence social time preferences in economic analysis.

As noted in the 2003 Green Book (and as described above), there are a range of estimates for the individual components of the discount rate. Researchers continue to illustrate a range of plausible estimates but conclude that the overall discount rate of 3.5% remains within this range and is justifiable [17].

In economics, the exponential utility discounted model [18] is the standard prescriptive model for social time preference [19]. The present value of a future outcome is calculated as expressed in the following formula:

$$PV = \frac{FV}{(1 + r)^t} \tag{8.3}$$

where:

- FV is the Future Value
- r is the discount rate
- t is the time period.

The discount rate is usually an annual rate, where t indicates the number of years before the future outcome is realized.

Conceptually, just as non-market goods and services can be assigned a monetary value, a discount rate can similarly be applied [20]. Therefore, just as compound investments grow exponentially over time, future values are discounted exponentially [21].

The present value of a future outcome can also be calculated through hyperbolic discounting, unlike exponential discounting (used in standard economics), hyperbolic discounting uses a hyperbolic function to model how people value future rewards. This function devalues the future less than an exponential function with the same discount rate:

$$\mathrm{PV} = \frac{FV}{1 + rt} \tag{8.4}$$

The use of this function devalues the future less than an exponential function (Rout e Walshe, 2013) is important because it more accurately reflects how people actually make decisions involving time, compared to traditional economic models. This insight has significant implications for economics, psychology, and public policy.

However, some studies show that people do not discount all outcomes at the same rate. For example, there is a tendency to discount gains at a higher rate than losses and large outcomes at a lower rate than small outcomes and [22]. Individuals may also apply different discount rates within different domains, for example, they may discount monetary outcomes, health outcomes, and environmental outcomes at different rates [21, 22].

Broome argues, however, that behavioral studies documenting contrasts in individuals' temporal preferences across different domains do not necessarily violate normative ideals of decision-making [23]. Furthermore, he emphasizes the theoretical validity of differential discounting of monetary and non-monetary consequences, in an attempt to reconcile purely material aspects (to which economists primarily refer) with well-being (on which environmentalists focus their interest), since the material outcomes investigated are results perceived only by the decision-maker, while outcomes related to well-being (e.g., environment and health) are experienced by the entire society. If these theoretical and behavioral perspectives are accepted, this domain independence has obvious implications for decision-making across multiple criteria.

Information on time preferences can be prescriptive, following the economic model with discount rates estimated from market interest rates, or descriptive, following a behavioral model with discount rates deduced from the decision-maker. Where the natural units associated with a criterion are non-monetary, a descriptive approach to time preference can be easily justified.

Here it is provided a comprehensive overview of social time preference and its application in decision-making processes, particularly in the context of public policy and infrastructure investments. It emerges that social time preference is a crucial concept in decision-making processes, particularly in public policy and infrastructure investments. This concept is embodied in the social discount rate, which plays a vital role in balancing short-term and long-term impacts of projects, especially those with environmental and social implications. The determination of an appropriate discount rate has been a subject of debate between economists and environmentalists, with the former typically advocating for higher rates that prioritize present benefits, while the latter push for lower rates to give more weight to future impacts. The UK Treasury's Social Time Preference Rate (STPR) serves as a practical example of how governments approach this issue, incorporating both "time preference" and "wealth effect" components. Various models of discounting, such as exponential and hyperbolic functions, have been proposed, each with distinct implications for valuing future outcomes. Interestingly, research has shown that individuals may apply different discount rates to various types of outcomes, a phenomenon known as domain-specific discounting. The field of social time preference also distinguishes between prescriptive approaches, which are based on market rates, and descriptive approaches, which are derived from observed behavior, in determining discount rates. This multifaceted approach to understanding and applying social time preference reflects the complexity of balancing present and future considerations in policy-making and investment decisions.

This text effectively illustrates the complexity of incorporating time preferences into multi-criteria decision analysis, especially for public policy decisions with long-term impacts. It underscores the need for a nuanced approach that can account for different types of outcomes and potentially varying discount rates across domains. The discussion touches on crucial debates in economics, environmental policy, and decision theory, highlighting the challenges of balancing present and future interests in a rigorous and justifiable manner. This is particularly relevant for issues like climate change, where the costs are immediate but many benefits are long-term and uncertain.

8.2.3 Methodology

After conducting a review of scientific literature to examine how time has been treated in various domains, we now illustrate how descriptive information on temporal preferences can be incorporated into a multi-criteria decision analysis, considering the peculiarities of the case study.

It is proposed a methodology that utilises a coefficient allowing for the comparison of heterogeneous performances occurring at different times and for different periods within the MCA, analogous to how the STPR defines temporal preferences in Cost–Benefit Analysis.

To this end, we introduce an intertemporal weighting coefficient tr that defines temporal preferences, just as criteria weights are a faithful representation of the decision-maker's preferences.

Each criterion is assigned its own τ_r, consistent with the STPR that normally applies to the domain to which the criterion belongs and specific to the year in which the evaluation is taking place.

Consistent with the above, τ_r is calculated in two ways:

- For criteria related to material or monetary aspects, an exponential rate of 3.5% is applied, adhering to Spackman's suggestion for the public sector [24], using the following formula:

$$\tau_r^e = \frac{1}{(1+0.035)^t} \tag{8.5}$$

  where τ_r^e represents the exponential intertemporal weighting coefficient and t is the time delay.
- For criteria related to the well-being of the entire society (social and environmental issues), a hyperbolic rate of 1.5% is applied as suggested by the Green Book, using the following formula:

$$\tau_r^h = \frac{1}{1 + 0.015t} \tag{8.6}$$

here τ_r^h indicates the hyperbolic intertemporal weighting coefficient and t is the time delay.

This approach allows for a nuanced treatment of temporal preferences within a multi-criteria framework, distinguishing between material/monetary criteria and those related to broader societal well-being. By applying different discount rates and functional forms (exponential for material aspects, hyperbolic for social and environmental issues), the methodology aims to capture the varying nature of temporal preferences across different domains.

The use of domain-specific temporal weighting coefficients aligns with recent research suggesting that individuals and societies may value future outcomes differently depending on the nature of those outcomes. This method provides a more flexible and potentially more accurate representation of societal preferences over time, particularly for long-term public policy and infrastructure decisions where environmental and social impacts play a significant role.

By integrating these temporal considerations directly into the multi-criteria analysis framework, the proposed methodology offers a promising approach for decision-makers to more comprehensively evaluate options that have diverse impacts occurring over different time horizons. This could be particularly valuable for sustainable development projects, where balancing short-term costs with long-term social and environmental benefits is often a key challenge.

It is possible to insert τ into the linear additive formula used to evaluate the alternatives and the evaluation is repeated for each year within the planned time horizon. The values, once homogenized in this manner, can finally be aggregated through summation into a matrix V of alternatives and actors, weighted by temporal preferences over the specified timeframe.

References

1. Macharis C, Bernardini A (2015) Reviewing the use of multi-criteria decision analysis for the evaluation of transport projects: time for a multi-actor approach. Transp policy 37:177–186
2. Frini A, Benamor S, Urli B (2020) Temporal MCDA methods for decision-making in sustainable development context. In: Sustainable concept in developing countries [Working Title], pp 1–14. https://doi.org/10.5772/intechopen.90698
3. Folkeson L, Antonson H, Helldin JO (2013) Planners' views on cumulative effects. A focus-group study concerning transport infrastructure planning in Sweden. Land Use Policy 30:243–253. https://doi.org/10.1016/J.LANDUSEPOL.2012.03.025
4. Zhou KY, Sheate WR (2011) Case studies: application of SEA in provincial level expressway infrastructure network planning in China—Current existing problems. Environ Impact Assess Rev 31:521–537. https://doi.org/10.1016/J.EIAR.2010.10.005
5. Dallocchio M (1995) Finanza d'azienda: analisi e valutazioni per le decisioni d'impresa. Egea
6. Gallo A (2014) A refresher on net present value. Harv Bus Rev 19
7. Dodgson J, Spackman M, Pearman A, Phillips L (2009) Multicriteria analysis: a manual

8. Betrie GD, Sadiq R, Morin KA, Tesfamariam S (2013) Selection of remedial alternatives for mine sites: a multicriteria decision analysis approach. J Environ Manag 119:36–46
9. Balana BB, Mathijs E, Muys B (2010) Assessing the sustainability of forest management: an application of multi-criteria decision analysis to community forests in northern Ethiopia. J Environ Manag 91:1294–1304
10. Scholten L, Schuwirth N, Reichert P, Lienert J (2015) Tackling uncertainty in multi-criteria decision analysis–an application to water supply infrastructure planning. Eur J Oper Res 242:243–260
11. Khalili NR, Duecker S (2013) Application of multi-criteria decision analysis in design of sustainable environmental management system framework. J Clean Prod 47:188–198
12. Walshe T (2013) Accounting for time preference in management decisions: an application to invasive species. J Multi-Criteria Decis Anal 211:197–211. https://doi.org/10.1002/mcda
13. Frini A, Benamor S (2018) Making decisions in a sustainable development context: a state-of-the-art survey and proposal of a multi-period single synthesizing criterion approach. Comput Econ 52:341–385
14. Banamar I, De SY (2018) An extension of PROMETHEE II to temporal evaluations. Int J Multicriteria Decis Mak 7:298–325
15. Meyer RF (1976) Preferences over time. In: Keeney RL, Raiffa H (eds) Decisions with multiple objectives: preferences and value tradeoffs
16. Layard R, Sargan J, Ager M, Jones D (1971) Qualified manpower and economic performance: an inter-plant study in the electrical engineering industry. Allen Lane
17. HM Treasury (2018) The green book central government guidance on appraisal and evaluation
18. Samuelson PA (1937) A note on measurement of utility. Rev Econ Stud 4:155–161
19. Frederick S, Loewenstein G, O'donoghue T (2002) Time discounting and time preference: a critical review. J Econ Lit 40:351–401
20. Rout TM, Walshe T (2013) Accounting for time preference in management decisions: an application to invasive species. J Multi-Criteria Decis Anal 20:197–211
21. Hardisty DJ, Weber EU (2009) Discounting future green: money versus the environment. J Exp Psychol Gen 138:329
22. Chapman GB (1996) Temporal discounting and utility for health and money. J Exp Psychol Learn Mem Cogn 22:771
23. Broome J (1994) Discounting the future. Philos Public Aff 23:128–156
24. Spackman M (2016) Appropriate time discounting in the public sector. Cent Clim Chang Econ Policy Work Pap

Chapter 9
The MACTOR Method and Its Case-Based Adaptation

Abstract This phase involves integrates the individual evaluations, carried out by each stakeholder, who has already individually expressed their criteria and weights based on temporal value. These evaluation are then combined into a single evaluation framework. The final matrix represents the synthesis of the evaluations developed thus far. Each element indicates the feasibility of a given alternative for a specific stakeholder based on their expressed preferences, strategic issues, and priorities. The chapter will be organised as follows. First, the MACTOR method and its case-based reworking will presents the analytical method developed by Godet. This method is based on an alliances and conflicts matrix and will be adapted and used to study the reciprocal influences among stakeholders, their strategies, their convergences and divergences in objectives, and the existing balance of power.

The MACTOR Method was developed to address the need for integrating individual evaluations into a comprehensive assessment framework, accommodating the various criteria and weights expressed by different stakeholders.

It becomes evident that in collaborative decision-making contexts, implementing any alternative proves relatively straightforward when stakeholders agree, yet nearly impossible when consensus is lacking. This challenge is further complicated by the disparate forms of decision-making power held by stakeholders, which they can leverage to advance their preferred alternatives and successfully execute their projects.

This analytical method, developed by Godet and based on a matrix of alliances and conflicts, enables the examination of reciprocal influences among stakeholders, their strategies, convergences and divergences in objectives, and existing power dynamics. This approach facilitates better utilization of the informational value generated through various working groups. Individual stakeholder inputs are transformed and analyzed to produce a range of outputs, examining convergences and divergences to ultimately develop a synthesis matrix [1].

It is therefore crucial to incorporate an analysis of potential alliances and conflicts into the evaluation process to more accurately represent reality. This is accomplished

E. M. Bertolini, *Temporal Value in MCDM*, Lecture Notes in Intelligent Transportation and Infrastructure, https://doi.org/10.1007/978-3-031-95485-6_9

through analyzing stakeholder moves, comparing their projects, and examining the power relationships between them.

To establish a consensus foundation for implementing optimal alternatives, Godet identifies two critical stages in the MACTOR method: first, a careful analysis of power dynamics among stakeholders—without which scenarios would lack relevance and coherence—and second, the development of a tool for systematic analysis of stakeholder behavior. The absence of a tool capable of identifying key variables is particularly limiting, as stakeholder behavior analysis is often preceded by inadequate structural analysis [1].

To focus on the key variables for future outcomes and the stakeholders who directly or indirectly control them, appropriate tools for identification must be employed. This involves asking the right questions to enhance the relevance of the evaluation process. Moreover, it is important to concentrate on those stakeholders who directly or indirectly control the key variables identified through structural analysis [2].

In this study, we employ a customized method for evaluating the reciprocal influence of stakeholders, whose general structure is adapted from the MACTOR method presented by Godet in 1991. Our approach represents a partial reworking of the original method; whereas Godet builds his reasoning on a property of binary matrices, our matrices are not binary. However, given their modest dimensions, we can achieve comparable results without rigidly adhering to the original methodology.

The following section provides a concise description of Godet's MACTOR method, while section "The MACTOR Method—Case Based," will detail the structure of the method as applied in our evaluation in section "Case-Based MACTOR Method."

9.1 The MACTOR Method

The MACTOR[1] method focuses on stakeholders who directly or indirectly influence the key variables identified through structural analysis. A "stakeholder strategies" table is constructed in the form of a square matrix (stakeholders × stakeholders) where:

- Each diagonal cell contains the aims and objectives of each stakeholder, to the extent these can be identified;
- The remaining cells contain the means of action that each stakeholder can employ against others to achieve their aims.

This table is developed through group discussions, based on comparing information collected about each stakeholder and their relationships with others. Information about stakeholder behavior can be gathered or supplemented through conversations

[1] As presented in the article Actors' Moves and Strategies: The MACTOR Method—An Air Transport Case Study by Michel Godet, published in the journal Futures (July/August 1991) by Butterworth-Heinemann Ltd (0016-3287/91/060605-18).

with experts representing each stakeholder group. Since stakeholders are typically reluctant to reveal their strategies and strengths/weaknesses, it is often more productive to ask for their opinions regarding other stakeholders' approaches. By carefully examining sets of partially accurate information, a relatively coherent picture of the entire situation emerges.

The goal is to create an analytical tool that maximizes the added value to be placed in the stakeholder strategy tables. Although game theory might seem appealing, Godet did not rigorously pursue this approach when creating MACTOR, instead prioritizing simplicity to ensure accessibility for users, particularly as it lends itself to multiple and varied applications. The analysis of stakeholder moves, as proposed by MACTOR. The following list explicates these stages as described by the author [1].

The distinctive value of the MACTOR method derives primarily from the phases of positioning stakeholders relative to their objectives and analyzing potential alliances and conflicts, from which strategic recommendations can be drawn.

1. Document the plans, motivations, constraints, and means of action of each stakeholder (construct the "stakeholder strategy" table).
2. Identify strategic issues and objectives associated with these battlefields.
3. Position each stakeholder on each battlefield and note convergences and divergences.
4. Rank objectives for each stakeholder and assess possible tactics.
5. Evaluate power relationships and formulate strategic recommendations for each stakeholder, aligned with the stakeholder's objective priorities and available resources.
6. Raise key questions about the future, formulating hypotheses about trends, events, and discontinuities that will characterize the evolution of power balances among stakeholders. Scenarios will be built around these key questions and hypothetical responses.

This schematic representation will be adapted to the case study for which we will adopt the Case-Based MACTOR method, the steps of which are described specifically in the following sections.

9.2 The Case-Based MACTOR Method

The Case-Based MACTOR method represents a reformulation of Godet's original MACTOR approach [1], specifically adapted for our case study. This modified methodology focuses on four key stakeholders and the strategic issues they identified during collaborative working sessions, derived from each stakeholder's individual criteria.

9.2.1 Analysis of Stakeholder Dynamics

Throughout the various working group meetings, stakeholder behaviors and strategic moves were studied to identify their objectives, examine power relationships in terms of constraints and means of action, and uncover key issues for future consideration that emerged from discussions.

Subsequently, through group deliberations, information about each stakeholder and their relationships with others was shared, enabling the development of a "stakeholder strategies" framework based on preferences revealed during the evaluation process. The level of analysis conducted thus far remains relatively basic, considering only these preferences. To enhance the model's alignment with reality, it is necessary to introduce two previously unconsidered dimensions:

- The strategic issues identified by stakeholders
- The implementation priorities for alternatives, which vary from one stakeholder to another.

This analysis will explicitly reveal the aims, objectives, and means that each stakeholder can employ to reach a common decision.

Identification of High-Value Alternatives for Stakeholders

Through the Multi-Criteria Analysis (MCA) conducted thus far, we have identified the alternatives to which individual stakeholders assign the highest value. We observe that these preferences significantly influence the dynamics of stakeholder interactions. For the following analysis, we will utilize the values calculated for the 2021–2051 timeframe.

Identification of Strategic Issues

The strategic issues relevant to each stakeholder have emerged quite clearly through examination of their planning and programming documents, as well as discussions during various working sessions. We note a significant correlation between these strategic issues and the criteria deemed most important by stakeholders. For simplification purposes, we will directly consider the latter.

For each criterion identifying strategic issues, only those alternatives that maximize and/or minimize the criterion will be selected. These alternatives will then be associated with the strategic issues.

Identification of Priorities

During the working sessions, each stakeholder was asked to rank alternatives according to their implementation priorities (if all alternatives could be implemented, in what order would they proceed?). Stakeholders numbered alternatives from 1 to 5, placing the highest priority alternative in position 1 and continuing in descending order.

Table 9.1 Compilation rules of the V matrix

Compilation rules of the V matrix	
+1	The alternative has the best performance for the stakeholder
0	The alternative has intermediate performance for the stakeholder
−1	The alternative has the worst performance for the stakeholder

9.2.2 *Stakeholder Positioning—Position Matrices*

Relationships between stakeholders can be represented as diagrams of potential alliances and conflicts. To comprehensively understand the situation, it is necessary to construct diagrams of possible convergences and divergences associated with each alternative in terms of preferences, strategic issues, and priorities.

Visual comparison of convergence and divergence diagrams is not immediately intuitive; however, a matrix representation allows us to summarize all these diagrams in a single table. For each stakeholder, the challenge is to identify and evaluate possible options and make a coherent selection of alliances, particularly because strategies of convergence and divergence among them vary depending on whether the focus is on alternative values, strategic issues, or priorities.

9.2.2.1 Matrix V (Value Matrix)

To understand convergences and divergences, only the alternatives that respectively have the best and worst performance for each stakeholder should be considered, based on the evaluation conducted thus far. In this case, matrix V (Value positioning matrix with sign, Stakeholders × Alternatives) is compiled according to the following rules (Table 9.1).

9.2.2.2 Matrix S (Strategic Issues Matrix)

Matrix S (positioning matrix with respect to strategic issues with sign) is compiled according to the following rules (Table 9.2).

Table 9.2 Compilation rules of the S matrix

Compilation rules of the S matrix	
+2	The alternative maximizes both criteria representing strategic issues for the stakeholder
+1	The alternative maximizes one criterion representing strategic issues for the stakeholder
0	The alternative has intermediate performance for the stakeholder
−1	The alternative minimizes one criterion representing strategic issues for the stakeholder
−2	The alternative minimizes both criteria representing strategic issues for the stakeholder

Table 9.3 Transformation of the priority scale expressed by the actors

Priority transformation scale					
Stakeholder-expressed priority order	1	2	3	4	5
Stakeholder positioning with respect to priorities	2	1	0	−1	−2

9.2.2.3 Matrix P (Priority Matrix)

To better highlight stakeholder positioning with respect to alternatives, we have chosen to modify the original 1–5 scale to a scale ranging from +2 to –2. The higher the absolute value, the more the stakeholder is interested in an alternative they consider a priority (Table 9.3).

This transformation yields a second position matrix of priorities (Stakeholders × Alternatives), which we will call matrix P, for which convergences and divergences will be sought.

9.2.3 Defining Convergence and Divergence

For each pair of stakeholders, it is valuable to identify the number of alternatives where they exhibit convergence or divergence regarding values, strategic issues, and priorities. This can be observed by examining matrices V, S, and P through pairwise comparison.

To construct the convergence matrix C (stakeholders × stakeholders), we identify the number of concordant factors for each stakeholder pair (two rows from matrices S, V, and P). These represent a convergence of interests, while discordant factors represent divergence.

The pairwise comparison for determining convergence is established through the following scoring system (Table 9.4).

By summing the scores obtained from comparing all alternatives for each stakeholder pair, we derive the convergence matrices Cv, Cs, and Cp.

The pairwise comparison for determining divergence follows this scoring system, with matrices Dv, Ds, and Dp compiled according to these rules (Table 9.5).

Table 9.4 Compilation rules of the C matrix

Compilation rules of the C matrix	
+1	Both stakeholders have the same positioning for a given alternative, whether positive or negative
+0.5	Both stakeholders have similar positioning for a given alternative (same sign but different values, or one neutral)
0	Both stakeholders have different positioning for a given alternative (opposite signs)

Table 9.5 Compilation rules of the D matrix

Compilation rules of the D matrix	
−1	Both stakeholders have different positioning for a given alternative (discordant values)
−0.5	Both stakeholders have dissimilar but not oppositional positioning for a given alternative (the alternative is indifferent to one but not to the other)
0	Both stakeholders have the same positioning for a given alternative (concordant factors), whether positive or negative

By summing the scores obtained from comparing all alternatives for each stakeholder pair, we derive the divergence matrices Dv, Ds, and Dp (stakeholders × stakeholders). Note that higher convergence corresponds to lower divergence and vice versa.

These matrices therefore indicate, for each stakeholder pair, the number of alternatives where they demonstrate convergence (Cn_{ij}) or divergence (Dn_{ij}).

9.2.4 Analysis of Power Dynamics

The interaction of potential convergences and divergences depends not only on the hierarchy of alternatives for each stakeholder but also on a stakeholder's capacity to impose their priorities and preferences on others—in other words, on power relationships.

Convergence or divergence between stakeholders alone is insufficient to actively advance or oppose an alternative. Resources are necessary to influence or shape the behavior of other stakeholders to one's advantage, requiring the development of a strategy to reach a favorable agreement.

To guide decision-making, it is useful to analyze power relationships through a matrix recording the direct and indirect influence of one stakeholder upon another. The influence matrix MI (stakeholders × stakeholders) is compiled according to the following scale (Table 9.6).

These aggregated values define the influence exerted and received, thus determining each stakeholder's level of dependence and independence relative to others. Specifically, this is accomplished by summing all values in a row of matrix I to

Table 9.6 Rules for completing the I matrix

0	No influence
1	Weak indirect influence
2	Weak Direct influence
3	Strong indirect influence
4	Strong direct influence

understand the level of independence (c_i), and all values in a column of the same matrix to understand the level of dependence.

We have seen that it is possible to account for each stakeholder's alternative priorities through matrix P (priority positioning matrix). Stating that one stakeholder is twice as influential as another in the general power equilibrium simply means attributing twice the influence of this stakeholder on the alternatives. It is therefore useful to describe power relationships between stakeholders using coefficients, allowing us to weight the rows of the priority evaluation matrix P with these coefficients.

These coefficients, to best represent power relationships, must consider both levels of independence and dependence. These two levels are calculated by aggregating the influence exerted with the influence received. To combine the values of dependence and independence, we first define a coefficient of relative independence (c_{ir}) that measures relative independence according to the following formula:

$$c_{ir} = \frac{c_i}{\sum c_i} \tag{9.1}$$

where:

c_{ir} is the coefficient of relative independence.

c_i is the independence value of the individual stakeholder.

It is then necessary to balance the previous coefficient c_{ir} with an inverse function of dependence, obtaining a second coefficient of relative independence and dependence (c_{idr}):

$$c_{idr} = \frac{c_i}{\sum c_i} * \frac{c_i}{c_i + d_i} \tag{9.2}$$

where:

c_{idr} is the coefficient of relative independence and dependence.

c_i is the independence value of the individual stakeholder.

d_i is the dependence value of the individual stakeholder.

Therefore, if dependence d_i is zero, $c_{idr} = c_{ir}$; if d_i, (the dependence) is strong relative to independence, then the power relationship c_{idr} will be much weaker than the simple relationship c_{ir}.

Furthermore, to facilitate understanding and subsequent calculations, Godet suggests introducing the power balancing coefficient $c*$ through the following formula:

$$c^* = \frac{c_i}{\sum c_{idr}} \tag{9.3}$$

where:

$c*$ is the power balancing coefficient.

c_{idr} is the coefficient of relative independence and dependence.

c_i is the independence value of the individual stakeholder.

9.2.5 *Stakeholder Positioning Matrices Balanced for Influence*

Having transformed power relationships between stakeholders into coefficients, it is now sufficient to weight the stakeholder positioning matrices (V, S, and P) with these coefficients.

We thus transition from positioning matrices V, S, and P to matrices V_i, S_i, and P_i—positioning matrices weighted by power relationships—by multiplying each row by the vector $c*$.

9.2.6 *Definition of Convergence and Divergence Balanced by Influence*

At this point, we can define new convergence and divergence matrices balanced by power relationships, respectively C_{si}, C_{vi}, C_{pi} and D_{si}, D_{vi}, D_{pi}. In this case, their compilation is less intuitive than the previous ones, though it still proceeds from pairwise comparison of the performance of all alternatives for all stakeholders. To define convergence and divergence, we calculate the distances between stakeholder positions through the difference of individual values in the pairwise comparison.

This reveals a direct proportionality between distance and divergence, and an inverse proportionality between distance and convergence: when distance is minimal, convergence will be maximal and divergence minimal, whereas when distance is maximal, convergence will be minimal and divergence maximal.

We have decided to normalize all obtained values to a single scale of divergence and convergence to enable more effective comparisons.

It is therefore possible to coherently compare the preferences of various stakeholders, considering both power relationships and preferences.

9.3 The Case-Based MACTOR Method

The construction of the decision matrix D is accomplished through a straightforward composition of the influence-weighted alternative/stakeholder matrices (Si, Vi, Pi) obtained previously. This matrix enables the classification of alternatives based on performance over time, strategic issues, and implementation priorities for each stakeholder.

Matrix D represents the synthesis of the evaluation conducted thus far, where each element dij indicates the feasibility of implementing a given alternative for a specific stakeholder as a function of expressed preferences, strategic issues, and priorities. The result obtained is balanced by power relationships and each stakeholder's margin for action.

$$D = V_i + S_i + P_i \tag{9.4}$$

where:

D is the decision matrix.

V_i is the matrix of alternative values positioning weighted by influence.

S_i is the matrix of strategic issues positioning weighted by influence.

P_i is the matrix of priorities positioning weighted by influence.

This integrated approach to decision matrix construction acknowledges the multi-dimensional nature of stakeholder evaluations. By combining these three critical dimensions—performance values, strategic considerations, and implementation priorities—while accounting for power dynamics, we achieve a more comprehensive and realistic assessment framework. This methodology enables decision-makers to navigate complex multi-stakeholder environments where power asymmetries and varying priorities significantly influence feasibility outcomes.

The resulting decision matrix provides a nuanced tool for identifying the most viable alternatives given the existing stakeholder landscape. It reflects not only technical performance but also political feasibility and strategic alignment, offering a holistic view that more closely mirrors real-world decision-making contexts.

References

1. Godet M (1991) Actors' moves and strategies: the mactor method. An air transport case study. Futures 23:605–622. https://doi.org/10.1016/0016-3287(91)90082-D
2. Godet M (2000) The art of scenarios and strategic planning: tools and pitfalls. Technol Forecast Soc Change 65:3–22

Chapter 10
Case Study: The Gorizia Interporto

Abstract This chapter provides a comprehensive analysis of the evaluation subject, with a focus on the Gorizia Interporto in the Sant'Andrea area. Situated within the broader logistics framework of Friuli Venezia Giulia, the terminal's strategic importance is assessed in terms of infrastructure, intermodal transport, and sustainable logistics. Five potential development alternatives are explored and evaluated for their impact on territorial cohesion, economic efficiency, and environmental sustainability. Through stakeholder analysis and multi-criteria assessment, the research examines the terminal's role in regional and international freight networks. This theoretical evaluation lays the groundwork for the practical application detailed in Chap. 11.

10.1 Theoretical Framework

In this study, the term "infrastructure" encompasses the fixed facilities essential for transportation operations, including roads, port facilities, railway tracks, airports, and pipelines. It also incorporates mobile assets, which consist of all vehicles utilized for transportation. In conjunction with transportation, the concept of logistics is integral to facilitating territorial reorganization and development.

A pertinent concept in this context is territorial logistics, defined as "the set of processes involving the planning, implementation, regulation, coordination, and promotion of logistics and transport infrastructures and activities that enable the orderly movement of people, goods, and information within a specific geographical area. The goal is to enhance territorial cohesion and accessibility to goods, services, individuals, and locations, both within and beyond the region" [1]. Consequently, decisions made within one sector of logistics invariably affect all other areas, creating mutual interdependencies.

Considering contemporary organizational and locational models of enterprises, along with advancements in transportation infrastructure and technology, the scope of logistics can be expanded from a territorial perspective [1]. Initially associated primarily with corporate production processes, logistics now both influences and

E. M. Bertolini, *Temporal Value in MCDM*, Lecture Notes in Intelligent Transportation and Infrastructure, https://doi.org/10.1007/978-3-031-95485-6_10

is influenced by mobility systems, infrastructure development policies, and, more broadly, spatial planning for territorial optimization [2].

From a mobility perspective, a territory functions as a complex system of socio-economic relationships, characterized by a dense network of "nodes" and "links." Within this framework, production and distribution plants, industrial complexes, ports, railway stations, airports, freight villages, and intermodal hubs are classified as "point infrastructures," whereas the transportation routes connecting them constitute "linear infrastructures."

The linear infrastructures considered in this study are those forming part of the Trans-European Transport Network (TEN-T), a strategic initiative aimed at enhancing communication routes for the efficient functioning of the European Union's internal market. The TEN-T is divided into two main components: the Core Network and the Comprehensive Network. The Core Network is structured around corridors (Core Network Corridors) designed to modernize infrastructure, facilitating passenger and freight transportation across the EU. The Comprehensive Network encompasses the infrastructure that connects all other European regions to the Core Network.

Among point infrastructures, a logistics hub serves as a key area for the movement and storage of goods, situated upstream of port terminals and integrated within an intermodal transport system. These hubs are equipped with warehouses, management services, information and telematics systems, and facilities where manufacturing activities can be conducted to transform semi-finished goods into final products. Essentially, a logistics hub represents an extension of the traditional freight village concept, providing a strategic means to optimize the transport-storage-distribution chain. It plays a pivotal role in ensuring effective market penetration at both national and international levels and varies in scale depending on the volume of goods handled.

A logistics platform fulfills two principal functions: (1) attracting and concentrating freight flows and (2) organizing intermodal transport chains by optimizing routes and deliveries. To effectively perform these roles, an optimal geographic location is essential, leveraging key transport axes and appropriate infrastructure. Furthermore, the platform must be designed as a high-quality industrial settlement for enterprises across various sectors, with an increasing emphasis on eco-sustainability. This association aligns logistics platforms with the concept of sustainable logistics [3].

Sustainable logistics is characterized by its focus on environmental and safety concerns alongside economic development needs. It integrates technologies, procedures, and activities aimed at minimizing the environmental impact of various segments of the logistics chain without compromising service quality or economic profitability. This approach can be conceptualized as a "sustainability triangle," comprising three core elements: economic efficiency, socio-territorial development, and the reduction of negative externalities. Key aspects of sustainable logistics include long-distance transportation, local distribution, intermodality, storage facilities, and packaging technologies.

10.2 Territorial Framework

Friuli Venezia Giulia (FVG) is an Italian region located in the northeasternmost part of the country, covering a total area of 7,845 km^2. It shares borders with Austria to the north (Carinthia—Hermagor and Villach-Land districts), Slovenia to the east (Upper Carniola, Slovenian Littoral, and Gorizia), Veneto to the west (Belluno and Veneto Orientale), and the Adriatic Sea to the south. Due to its central position relative to Central Europe, Friuli Venezia Giulia has historically functioned as a key logistics platform in the Mediterranean, serving as a meeting point for multimodal corridors and a gateway to the Balkans, as well as a strategic node connecting Europe and the East.

In the context of this research, which focuses on the transportation system of Friuli Venezia Giulia, particular attention is given to two corridors within the TEN-T network: the Mediterranean Corridor, which accommodates most transit vehicles traveling between Austria and Slovenia, and the Baltic-Adriatic Corridor, which links northeastern Italian ports with Poland's Port of Gdańsk, via Austria, the Czech Republic, and Slovakia.

The region's transportation infrastructure includes an extensive motorway network (236.6 km with six main routes) and a primary road system (3,200 km of state, regional, and former provincial roads). Regarding rail transport, the region has 466 km of operational tracks, of which 341 km are primary lines and 125 km are complementary lines. The network consists of 299 km of double-track lines and 167 km of single-track lines, with 382 km electrified and 84 km diesel-powered. A key element of intermodal transport within the corridors is the Adriatic port system, particularly the ports of Trieste and Venice, which are adjacent to Northeast Italy's railway axes.

The port infrastructure in Friuli Venezia Giulia encompasses the ports of Trieste, Monfalcone, and Porto Nogaro. Among these, the Port of Trieste serves as the principal hub of the Eastern Adriatic Port System, which also includes Monfalcone. Trieste is a significant international port, facilitated by its railway connections to European production centers, and functions as a gateway for trade with the Far East and intra-Mediterranean markets. Notably, the Free Port of Trieste operates under a distinctive legal framework, setting it apart from other free zones. The Port of Monfalcone, recognized as a port of national interest, is the northernmost Mediterranean port specializing in general and dry bulk cargo. In contrast, Porto Nogaro, designated as a regional port, is situated on the Corno River and accessed via a three-mile translagoonal channel.

The 2015 port reform resulted in the creation of the Eastern Adriatic Port System Authority (AdSP), which aims to achieve functional and operational integration at the regional level. The AdSP is also tasked with coordinating interport and hinterland logistics systems.

Friuli Venezia Giulia hosts four principal logistics hubs: the Fernetti Freight Village, the Sant'Andrea Freight Village (the focus of this case study), the Cervignano Freight Village, and the Pordenone Freight Village.

Fernetti Interport is located in the municipality of Monrupino, near the Italian-Slovenian border, and serves as a crucial retroport facility. It provides essential support to the maritime terminals of Trieste and Monfalcone through integrated road and railway connections, functioning as a critical interface between maritime and terrestrial transportation systems.

Sant'Andrea Interport will be examined in detail in a subsequent section of the analysis.

Cervignano Interport is situated adjacent to a railway junction and represents a strategically positioned intermodal terminal. Covering an area of 160,000 m^2, it features three railway tracks, each 750 linear meters in length. The facility is equipped with a road weighbridge and approximately 17,000 m^2 of covered storage space, including two overhead cranes with lifting capacities of 12.50 and 30 tons. Its geographical location is particularly advantageous, with proximity to key regional ports: Porto Nogaro (11 km), Monfalcone Port (29 km), and Trieste Port (48 km).

The Pordenone Wholesale Logistics Center is located near Pordenone's urban area and is the sole active wholesale logistics and commerce center in the Friuli Venezia Giulia region. It benefits from a railway connection integrating the Venice-Udine line and features an independent highway exit on the A28 Portogruaro-Conegliano motorway.

The region is also served by an airport located in Ronchi dei Legionari, which is directly connected to the railway network and bus station, thereby enhancing its multimodal transportation capabilities.

Each of these facilities plays a pivotal role in regional and international freight movements, capitalizing on their strategic locations and intermodal capabilities.

The region's transport infrastructure comprises an intricate network that includes maritime ports, road and rail networks, border autoports, and logistics centers. This integrated transportation system represents a valuable asset that necessitates coordinated management to ensure functional alignment with EU and national policies on intermodality and logistics innovation (Trieste Report, 2003—Association of Industrialists of Trieste). The regional administration also acknowledges the critical importance of its development.

The Regional Government of Friuli Venezia Giulia has adopted a strategic approach to logistics development. In its resolution of June 29, 2007 (No. 1592), it emphasized the necessity of establishing a comprehensive transport system that integrates infrastructure, services, and logistics networks into a unified framework. The document conclusively underscores the paramount objective of establishing a comprehensive "Transportation System" conceptualized as a cohesive network of infrastructure and service offerings.

The ultimate vision is to create an integrated, efficient, and adaptive logistics ecosystem that can effectively support regional economic growth and international trade dynamics.

10.3 The Evaluation Subject in Gorizia

The Sant'Andrea Interporto is situated in the province of Gorizia, within the Isontino region. This area, named after the Isonzo River that traverses it, encompasses most of the province of Gorizia. It is bordered to the northwest by the Collio hills, to the east by the initial elevations of the Karst Plateau, and to the west by the Judrio River, which marks the boundary with the province of Udine. The territory extends southward to the Adriatic Sea, encompassing Grado and its lagoon. The province benefits from an extensive and well-developed transport network, including roads, highways, airports, and maritime routes, ensuring high accessibility. The road infrastructure is particularly dense, providing efficient connectivity. The principal road arteries include State Roads No. 14 (Venezia Giulia), No. 55 (Isonzo), No. 56 (Gorizia), No. 252 (Palmanova), No. 305 (Redipuglia), No. 351 (Cervignano), No. 352 (Grado), No. 356 (Cividale), and No. 409 (Plessiva), alongside the A4 Torino-Trieste motorway. The railway network consists of the Monfalcone-Udine and Venice-Villa Opicina lines. The presence of the port, the Trieste-Udine and Trieste-Venice railways, and the Ronchi dei Legionari Airport makes Monfalcone a crucial transit hub, serving as a gateway between Trieste, Isontino, and Friuli and a key point for international connections with Central and Eastern Europe.

The Gorizia Interporto is an extensive infrastructure complex designed to facilitate the exchange of goods across various modes of transport. It is one of the seven logistics hubs constituting the Friuli Venezia Giulia platform, which includes three maritime ports (Trieste, Monfalcone, and Porto Nogaro) and four inland terminals (Isontino Terminal, Fernetti, Cervignano, and Pordenone).

The terminal comprises the following facilities:

- A logistics platform that includes warehouses for freight forwarders, couriers, and goods management operators.
- A railway yard capable of assembling complete trains and connecting with other ports and inland terminals across the continent.
- General support services, such as banking, postal services, restaurants, and fuel stations.
- Specialized services, including customs and telematics facilities.
- Parking areas.

The total area of the terminal covers 600,000 m^2, including:

- 43,000 m^2 of covered areas,
- 260,000 m^2 of equipped areas,
- 20,000 m^2 allocated to the intermodal terminal,
- and 33,000 m^3 of refrigerated storage.

Additionally, the terminal is equipped with a photovoltaic system consisting of 6,640 panels covering an area of 20,000 m^2, generating over 1 MW of power.

10.3.1 Historical and Strategic Context

The terminal was developed in accordance with the strategic directives, policies, and financial provisions established under the "Osimo Agreements," signed in 1975 between Italy and the former Yugoslavia. These agreements were ratified on March 14, 1977 (Law No. 73/77) and came into force on October 11, 1977, subsequently formalized under Presidential Decree No. 100 of March 6, 1978. As part of these agreements, infrastructure projects were initiated to enhance highway connections between Italy and Yugoslavia via the Fernetti, Rabuiese, and Pesek border crossings. Consequently, the Gorizia Interporto was established as an intermodal logistics platform, offering a wide range of services for international trade and road-rail transport to and from Central and Eastern Europe.

In 1982, the Autoport Customs Stations of Gorizia, known as SDAG, were established in successive phases. Following Slovenia and Croatia's independence from the Yugoslav federation, Italy quickly recognized these nations as the legitimate successors to Yugoslavia's international commitments, including the Osimo Treaty.

The terminal's role evolved significantly after Slovenia joined the European Union in 2004, during a major phase of EU enlargement. Over its thirty-year existence, the terminal developed infrastructure for warehousing, office space, craft enterprises, parking, road connections, a railway junction, animal transit facilities, and customs and freight forwarding services. However, with the abolition of customs operations, the terminal underwent functional transformations, particularly in border-related activities. On April 27, 2010, the Municipality of Gorizia, which owns the terminal facilities, became the sole shareholder of SDAG, making it an in-house company under municipal control.

10.3.2 Strategic Position and Current Operations

The terminal is strategically located at the intersection of major transportation corridors between Central Europe and the Upper Adriatic port system. It benefits from its integration into the Trans-European Transport Network (TEN-T) corridors and is positioned at the Italy-Slovenia border, connecting the Italian A34 motorway—an extension of the A4 (European Route E70) at the Villesse junction—with the Slovenian H4 motorway, which leads to the A1 towards Ljubljana.

Currently, the terminal's primary operational sectors encompass the following:

- Logistics for fresh and frozen food,
- Parking and truck movement (with 80,000 truck entries recorded last year),
- Terminal concessions (providing 550 railway terminal slots and warehouse leasing), generating an annual revenue of approximately two million euros.

The terminal offers a comprehensive array of services for international trade and transportation, particularly for road and rail shipments to and from Central and Eastern Europe. It functions as a hub for cargo storage and consolidation, managing dry, perishable, fresh, and frozen goods, including those necessitating customs procedures. The terminal's competitive edge is derived from its specialized agri-food logistics operations, branded as GOFOODLOG (Gorizia Food Logistics). A recently completed agri-food platform focuses on fresh and frozen goods, featuring 11 modular cold storage units with adjustable temperatures ranging from +18 to −28 °C, enabling precise thermal tracking of stored products.

The terminal's strategic location facilitates swift access to Central and Eastern European markets, fostering the establishment of 50 businesses. It serves as a vital gateway for logistics routes towards Ljubljana, Hungary, and Eastern Europe. Following Slovenia's accession to the Schengen Area in 2007 and the subsequent elimination of border controls, the terminal experienced a resurgence after a period of stagnation. Prior to December 22, 2007, legal entry into Slovenia or Italy required identity verification at border crossings, where Italian and Slovenian authorities conducted customs inspections. Despite the removal of these controls, Gorizia's advantageous geographic position remains a critical asset, bolstered by ongoing infrastructure enhancements, such as the completion of highway links to the Italian A4 and Slovenia's Ljubljana-Koper motorway, as well as its proximity to urban centers like Gorizia, Nova Gorica, Sempeter-Vrtojba, the Upper Adriatic port system, and Ronchi dei Legionari Airport.

The terminal's growing significance is underscored by the sale of 73,700 Slovenian toll stickers (vignettes) by SDAG, significantly surpassing the 22,750 Hungarian vignettes sold, reflecting the terminal's increasing role in cross-border transport and logistics.

10.4 Stakeholder Analysis

For the purposes of this study, stakeholders in major sustainable infrastructure projects are defined as "those who can influence the project process and/or final outcomes, whose living environments are positively or negatively affected by the project, and who receive associated direct and indirect benefits and/or losses" [4]. In pursuit of achieving the most realistic possible perspective on the ongoing dynamics, and in alignment with current transportation and planning regulations, the key entities involved in the assessment are identified as follows.

10.4.1 SDAG: A Management Entity Responsible for the Border Station and Terminal of Sant'Andrea

The Stazioni Doganali Autoportuali di Gorizia—SDAG (Gorizia Border Station Authority) organization is entrusted with the following critical responsibilities.

The comprehensive management, maintenance, and operational oversight of the border station and terminal infrastructure within the Municipality of Gorizia—specifically in the Sant'Andrea locality—including all associated services. This mandate extends to the implementation of infrastructural, technical, and systems-related interventions, whether executed directly or through concession.

Furthermore, SDAG is commissioned to conduct independent research, feasibility studies, and strategic investigations. These activities are undertaken either on behalf of the Municipality of Gorizia or in conjunction with national and European projects, with the primary objectives of fostering territorial economic growth and development.

A key strategic focus is the promotion of research and development settlements dedicated to innovation and entrepreneurship, ultimately aimed at enhancing the region's economic valorization. By supporting emerging enterprises and innovative ecosystems, SDAG plays a pivotal role in driving local economic transformation and competitiveness.

10.4.2 Municipality of Gorizia: A Comprehensive Territorial Governance Framework

The Municipality of Gorizia serves as both the territorial jurisdiction hosting the Sant'Andrea Interport and the primary property owner of the infrastructure. Its territorial planning function is executed through a sophisticated governance approach that meticulously adheres to fundamental principles of appropriateness, regional interest, and subsidiarity.

The municipality has planning mandate, but it need to be aligne with the Territory Governance Plan (Piano di Governo del Territorio, PGT), maintaining a delicate balance between local autonomy and broader regional strategic objectives.

Leveraging the principles of subsidiarity and appropriateness, the municipality extends its territorial planning competence to the supra-municipal level when local-level objectives cannot be adequately achieved. This flexible governance model allows for strategic interventions that transcend traditional municipal boundaries, enabling more comprehensive and effective territorial development.

A critical dimension of the municipality's strategic positioning is its sole ownership of SDAG, effectively transforming the organization into an in-house entity as of April 27, 2010. This direct administrative control provides a powerful mechanism for implementing integrated territorial development strategies, ensuring close alignment between municipal objectives and operational implementation.

10.4.3 Eastern Adriatic Sea Port Authority: Strategic Governance and Regulatory Framework

The Eastern Adriatic Sea Port Authority represents a sophisticated public administrative entity with a critical mandate to coordinate and strategically manage maritime infrastructure development, operationally grounded in Law No. 84 of January 28, 1994.

Structurally, the Authority is configured as a non-economic public institution, distinguished by comprehensive administrative, budgetary, and financial autonomy. This institutional design is explicitly defined in Article 6, Paragraph 5 of the referenced legislation, as subsequently modified by Legislative Decree No. 169 of August 4, 2016. The organization is subject to direct supervisory oversight by the Minister of Infrastructure and Transportation.

The Authority has been entrusted with a broad spectrum of administrative functions, several of which were previously executed by maritime regulatory bodies. This expanded competency reflects an evolutionary approach to maritime governance, centralizing strategic coordination and operational management.

Financial governance represents a particularly notable aspect of the Authority's operational framework. Specifically, fiscal deliberations—encompassing both prospective and retrospective budgetary documentation—are subject to joint ministerial scrutiny. This dual-ministerial review process, involving the Minister of Infrastructure and Transportation in concert with the Minister of Economy and Finance, ensures rigorous financial transparency and accountability.

10.4.4 The Autonomous Region of Friuli Venezia Giulia

It represents a sophisticated administrative entity exercising comprehensive governance functions across urban planning and territorial development domains. Through strategic legislative frameworks, the region has established a robust mechanism for comprehensive territorial management.

Specifically, the region has implemented two critical legislative instruments to govern its administrative scope. Regional Law No. 5, enacted on February 23, 2007, provides a nuanced regulatory framework for urban development and territorial planning. Complementarily, Regional Law No. 23, dated August 20, 2007, addresses a wide spectrum of transportation-related governance, encompassing regional and local public transportation systems, freight logistics, vehicle motorization, road circulation, and infrastructure development.

Within these legislative domains, the region systematically executes a multifaceted administrative approach. This approach integrates sophisticated functions of strategic planning, programmatic development, directional guidance, systemic coordination, continuous monitoring, and rigorous regulatory surveillance.

In the subsequent analytical chapters, a systematic abbreviation protocol has been established to ensure concise and consistent referencing of key stakeholder entities involved in the infrastructural and territorial development analysis. This notation system facilitates streamlined communication while maintaining precise institutional identification.

Abbreviation Compendium

SDAG Stazioni Doganali Autoportuali di Gorizia (Gorizia Border Station Authority)
GO Municipality of Gorizia
ASP Autorità di Sistema Portuale del Mare Adriatico Orientale (Eastern Adriatic Sea Port System Authority)
FVG Autonomous Region of Friuli Venezia Giulia

The deployed abbreviations are derived from the primary institutional designations, for enhance textual readability while preserving precise institutional referencing. Each abbreviation corresponds directly to the full organizational nomenclature, ensuring unambiguous interpretation throughout the subsequent analytical discourse.

Readers are advised to reference this notation guide when encountering these abbreviated institutional references in the subsequent chapters to maintain interpretative clarity and conceptual precision.

10.5 Development Alternatives Analysis

Considering the strategic geographical positioning of the Isonzo node, which enables service provision along both the north–south axis (connecting Tarvisio-Udine-Gorizia-Trieste) and the east–west corridor (Gorizia-Udine-Venice and Gorizia-Monfalcone-Venice routes), and its proximity to the Italian-Slovenian border, several development alternatives for the SDAG platform emerge as potential candidates for multi-criteria assessment:

- Alternative Zero: Maintenance of Current Status
- GoFoodLog: Agrifood Asset Development
- Safe and Secure Park: Establishing Secure Truck Parking Zones
- Simplified Logistics Zone
- Intermodal Terminal

The subsequent research provides a detailed description of the alternatives considered in the Multi-Criteria Analysis (MCA), which constitutes the core focus of this investigative study.

Alternative Zero: Maintenance of Current Status

This alternative is comprehensively addressed in the dedicated section of the research, which provides a comprehensive examination of the existing infrastructural and operational conditions.

GoFoodLog: Agrifood Asset Development

The GoFoodLog alternative represents a transformative strategic intervention designed to reposition the SDAG platform as a pivotal nexus for advanced agrifood logistics within the Friuli Venezia Giulia region. This comprehensive approach aims to establish a sophisticated, integrated logistics infrastructure capable of providing holistic solutions for the complex management of agricultural and food-related products.

This agrifood asset represents a critical functional component within the regional logistical ecosystem, characterized by a coordinated network of ports and intermodal terminals. The proposed integrated offering is designed to provide sophisticated, adaptable solutions that effectively meet the complex operational demands of both national and international stakeholders in the agrifood sector.

The proposed model demonstrates a understanding of contemporary logistics challenges, offering a flexible, adaptive framework that can respond to the strains of global agricultural and food product distribution networks. Through this strategic reimagining of logistics infrastructure, the GoFoodLog alternative presents a compelling vision for regional economic innovation and competitive positioning.

Safe and Secure Park: Strategic Development of Secure Truck Parking Facilities

The Safe and Secure Park alternative represents a comprehensive strategic intervention addressing critical infrastructure challenges in transportation logistics. This proposed development responds to the multifaceted imperative of establishing secure and protected parking areas for commercial vehicles, with far-reaching implications for transportation safety, driver welfare, and regional economic development.

The primary objective of this alternative is to mitigate systemic vulnerabilities in current truck parking infrastructure by implementing advanced security protocols and service enhancements. Specifically, the initiative targets the comprehensive revitalization of service provisions for transportation professionals at the Sant'Andrea Station, contextualizing the intervention within the broader transformative landscape of European Union expansion and the implementation of the New European Customs Code.

The development is particularly significant in light of recent European Union regulatory frameworks. A standardized EU-level approach to truck parking facilities will generate multiple systemic benefits, including enhanced transparency for transportation professionals and creating a more conducive environment for private sector investment. Such standardization represents a critical step towards harmonizing transportation infrastructure across European jurisdictions, facilitating more efficient and secure cross-border logistics operations.

By integrating advanced security technologies, comprehensive service amenities, and regulatory compliance mechanisms, the Safe and Secure Park alternative presents a holistic approach to addressing contemporary challenges in transportation logistics infrastructure.

Simplified Logistics Zone: Strategic Economic Development Framework

The proposition to establish a Simplified Logistics Zone (hereinafter referred to as ZLS) represents a sophisticated policy intervention with significant potential for regional economic transformation. This alternative emerges as a strategically nuanced approach to institutional development, specifically designed to cultivate a preferential regulatory environment that systematically promotes economic activities intrinsically linked to maritime and logistics infrastructure.

Conceptually, the Simplified Logistics Zone paradigm represents the most contemporary manifestation of advanced institutional policy thinking regarding regional economic development. The strategic framework is constructed to leverage port-centric ecosystems, with a primary objective of catalyzing entrepreneurial settlements, implementing targeted economic incentives, and generating investment projects capable of propelling key economic sectors.

The proposed ZLS model demonstrates a comprehensive approach to sectoral development, strategically targeting high-value industries that constitute the backbone of contemporary economic competitiveness. These priority sectors include the sophisticated agricultural and food processing industries, the technologically advanced aerospace sector, the dynamic automotive manufacturing ecosystem, and the broader constellation of industries embodying the quintessential Italian design and production ethos—collectively encapsulated by the internationally recognized "Made in Italy" brand.

By creating a specialized economic zone with streamlined regulatory mechanisms, the ZLS alternative offers a nuanced policy instrument for stimulating economic innovation, attracting strategic investments, and reinforcing the region's competitive positioning within both national and international economic landscapes. The approach transcends traditional economic development strategies by providing a holistic, integrated framework that simultaneously addresses infrastructural, regulatory, and sectoral development imperatives.

Intermodal Terminal: Strategic Infrastructural Development and Cross-Border Connectivity

The proposed intermodal terminal alternative represents a sophisticated infrastructural intervention strategically designed to maximize the extraordinary geographical potential of the terminal's location. This unique positioning distinguishes the logistics hub as a critical nexus, situated not merely adjacent to a significant highway axis, but precisely at the geopolitical interface between Western and Eastern European transportation networks.

The terminal development emerged from a planned transnational infrastructure project, substantiated through financial investments from two pivotal European cooperative programs[1]: the Interreg III A Italia-Slovenia 2000–2006 initiative and the DOCUP Objective No. 2—2000–2006 program. The comprehensive investment, amounting to approximately €5,500,000, facilitated the construction of sophisticated logistics structures specifically designed for modal interchange between road and rail transportation systems.

The project's strategic location at the San Andrea Border Station—mere meters from the Slovenian border—underscores its geopolitical significance. The TIP Project identified two critical interventions essential for enhancing connectivity and accessibility along the international railway corridor connecting Gorizia Centrale and Nova Gorica. These interventions specifically focused on developing two specialized railway connectors, referred to as "lunettes," which ingeniously link the international Gorizia-Vrtojba-Nova Gorica line (upon which the SDAG intermodal system is integrated) with national railway lines traversing both Italian (Trieste-Udine) and Slovenian (Jesenice-Sežana) territories.

10.6 Contextualizing the Broader Cooperative Framework

In the subsequent analytical chapters, a systematic abbreviation protocol has been established to ensure concise and consistent referencing of the evaluated development alternatives. This notation system facilitates streamlined communication while maintaining precise conceptual identification of the proposed strategic options.

Alternative Abbreviation Compendium

A0	Alternative Zero (Maintaining Current Operational Status)
A GFL	GoFoodLog Alternative (Agri-Food Logistics Development)
A SSP	Safe and Secure Park Alternative (Secure Truck Parking Infrastructure)
A ZLS	Simplified Logistics Zone Alternative
ATI	Intermodal Terminal Alternative

The deployed abbreviations follow a structured approach that combines:

- An initial 'A' denoting 'Alternative'
- A concise acronym or abbreviation representing the specific development strategy

[1] The European Commission's proactive approach to transfrontier cooperation between Italy and Slovenia manifested through strategic co-financing mechanisms during the 2000–2006 period. The Interreg III A Community Initiative specifically targeted maritime and terrestrial border regions, encompassing Italian provinces including Udine, Gorizia, Trieste, Venezia, and to a lesser extent, Pordenone and Rovigo. The Slovenian regions of Obalno-Kraška, Goriška, and the municipality of Kranjska Gora were correspondingly integrated into this collaborative framework.

References

1. Iannone F, Aponte D (2006) La pianificazione e la promozione di un sistema logistico territoriale: linee guida ed esperienze nazionali ed europee. In: Atti della XXVII Conferenza scientifica annuale AISRe, Pisa, pp 12–14
2. Iannone F (2009) Analisi multicriteria per la classificazione di possibili interventi di potenziamento logistico dell'Area vasta pometina secondo le dimensioni ACIT. Working paper SIET 2009, ISSN 1973-3208
3. Iannone F (2003) Aspetti pubblici e privati di un modello di logistica sostenibile. Atti della VI Riun Sci Annu della Soc Ital degli Econ dei Trasp
4. Li THY, Ng ST, Skitmore M (2012) Public participation in infrastructure and construction projects in China: from an EIA-based to a whole-cycle process. Habitat Int 36:47–56

Chapter 11
Application of the Methodology and Performance Analysis

Abstract This chapter applies the methodology developed in the preceding chapter (Chap. 9), as specifically formulated through the combination of several existing MCA approaches. Building upon the methodological considerations summarised in Chap. 1: *The Problem of Choice*, this methodology has been adapted fot the case study under evaluation. For a detailed examination, please refer to Chap. 10: *The Case Study. The Gorizia Interporto.* The discussion begins with a description of the working group and continues with an exposition the economic and welfare criteria adopted by the stakeholders, as well as the determination of various weights. Initial rankings will be compiled, showin the short-, medium-, and long-term performance of the alternatives separately for each stakeholder. These rankings will then be transformed into multi-period assessments by aggregating the values obtained into a single measure that also takes into account how long the effects would persist across various periods. The analysis will then proceed with an examination of possible alliances and conflicts among stakeholders, developing a matrix that records the influence of one stakeholder upon another. This will enable a decision matrix to be compliled that considers not only individual preferences but also the roles of stakeholders and their potential courses of action. A key feature of this evaluation model was stakeholders involvement, which constituted what we will call the working group. The complexity of the topics and information addressed required active and collective participation to ensure greater objectivity in the assessments. The composition of this group was fundamental to the research, and the limited number of participants facilitated interaction and enhanced synergy.

11.1 Defining Criteria for Stakeholders

This section outlines the sustainability criteria selected to reflect the objectives of each stakeholder. During one of the initial GLD meetings, each stakeholder identified their specific criteria, which subsequently formed the basis for evaluating alternatives.

E. M. Bertolini, *Temporal Value in MCDM*, Lecture Notes in Intelligent Transportation and Infrastructure, https://doi.org/10.1007/978-3-031-95485-6_11

11.1.1 SDAG Criteria

SDAG has identified the following criteria:

- Achievement of budget balance
- Opportunity to obtain national or EU economic incentives
- Overcoming local marginality
- Expansion of SDAG activities
- Attraction of new businesses and clients to SDAG
- Economic development of the Gorizia area.

These criteria are based on SDAG's Multi-year Programming Document 2015–2022 (rev. 02 October 2018) and the Management Report on the financial statements as of December 31, 2017.

Achievement of Budget Balance

The financial statement represents the primary source of information for understanding an enterprise's financial health at a specific moment, typically coinciding with the end of the administrative period called the fiscal year. Within a financial statement, data is divided into two accounting statements: the balance sheet and the income statement.

For SDAG, as for any other company, a balanced budget—the equality of totals in the two sections of an account or financial statement—occurs when, during a year, financial outflows equal the inflows achieved, avoiding deficit situations and consequent recourse to debt or monetization.

Opportunity to Obtain National or EU Economic Incentives

The opportunity to obtain national economic incentives falls under the purview of the Ministry of Economic Development, in conjunction with the Ministry of Labor and Social Policies and Cassa Depositi e Prestiti. Various funding programs aim to facilitate business access to credit and enhance the country's productive system competitiveness, promote the diffusion and strengthening of the social economy by supporting the establishment and growth of enterprises operating throughout the national territory for the pursuit of worthy general interests and social utility purposes.

Regarding the opportunity to obtain EU economic incentives, the EU finances various projects and programs, applying strict rules to ensure rigorous control over the use of funds. Funding opportunities are numerous and of various types (by heading, cluster, and program or fund). The 2021–2027 multiannual financial framework and NextGenerationEU provide an overview of funding opportunities. Of particular interest for this criterion, under Heading 1 (Single Market, Innovation, and Digital Agenda), in the "European Strategic Investment Program" section there are three projects to which SDAG could contribute:

Connecting Europe Facility, which aims to support the creation of essential energy, transport, and digital infrastructures throughout Europe.

Digital Europe Program, the first EU program aimed at accelerating recovery and guiding Europe's digital transformation.

Single Market Program, which aims to protect and strengthen consumers' position, ensure food safety, and enable EU small and medium-sized enterprises to thrive.

Overcoming SDAG's Local Marginality

According to the criterion of overcoming local marginality, it is necessary to reintegrate SDAG into regional and national contexts. This condition can be overcome by leveraging territorial resources linked to its strategic location at the intersection of various routes to and from Central Europe. Alternatives should be aimed at improving transit service offerings to attract the flow directed to Eastern Europe. The same applies to logistics operations, enabling SDAG to express its excellence relative to competition through the development of information and telecommunications technologies for advanced tertiary services.

Expansion of SDAG Activities

The expansion of SDAG activities consists of broadening and diversifying the offerings that can satisfy the growing freight traffic; it is correlated with logistics development and increased productivity and efficiency of transport activities. It is oriented toward revitalizing and enhancing the system of services offered to users, such as selling Hungarian electronic vignettes, onboard devices for heavy vehicles, and tourist materials.

Attraction of New Clients to SDAG

It is necessary to design and implement actions that encourage new users to choose SDAG spaces by offering a wide range of services, such as information, sale of products and services related to both tourist and truck transit, and providing a service level adequate to target expectations. Alternatives will be evaluated based on their ability to attract an increasingly numerous user base.

Economic Development of the Gorizia Area

According to the criterion of economic development of the Gorizia area, it is necessary to overcome marginality. Urban marginality is understood as a problem associated with lack of opportunities, characterized by a relatively low level of socioeconomic development, demographic decline, and crisis in productive activities. This condition can be overcome by leveraging territorial resources linked to strategic location, at the intersection of various routes to and from Central Europe. Alternatives can initiate mechanisms to overcome local marginality, capable of reactivating development processes in cross-border areas, even within a new vision and public–private partnership. Actions aimed at requalifying human resources and implementing interventions that can combine socially-oriented action logic with business culture are fundamental. Relevant factors for development processes also include local leadership, social capital, and diversification of local economies.

11.1.2 *Municipality of Gorizia Criteria*

The Municipality of Gorizia has identified the following criteria:

- Economic and entrepreneurial development
- Community development
- Services for citizens.

The description of these criteria refers to the revision documents of the current Municipal General Regulatory Plan, Municipal Council Resolution No. 69 of 2019: Socio-Economic Territorial Study preliminary to the revision of the Gorizia Municipal General Regulatory Plan elaborated in February 2020; and to the Local Agenda 21 document, which represents an instrument through which the Administration defines not only its environmental protection and socioeconomic development objectives but also, and especially, the strategies and actions to implement them.

Economic and Entrepreneurial Development

The Municipality of Gorizia aims to achieve economic and entrepreneurial development by integrating and balancing three dimensions: Environment, Economy, and Society—that is, pursuing the objective of Sustainable Development.

To concretely apply a Sustainable Development policy in all its dimensions (social, economic, environmental, institutional), the Municipality intends to evaluate alternatives based on their contribution to the following actions and interventions: strengthening market instruments, redistributing resources, and implementing institutional and economic reforms. From this perspective, an indispensable integrated economy will be established, promoting the creation of new businesses and employment.

The revitalization of Gorizia can be articulated through the sectors of innovative industry, craftsmanship, tourism, commerce, and agriculture; in this way, the territory can benefit from greater integration, including cross-border integration, from a logistical perspective. The identification and strengthening of the commercial circuit can constitute the driving factor of integrated economic development if promoted in sustainable terms and reconciled with the quality of life of the resident population. The prospect of strengthening a commercial circuit that functions as a natural shopping center, with adequate interventions on mobility (access to the area through new roads and parking), will lead to an improvement in urban quality (redevelopment of public spaces that can compete with the "new squares" of shopping centers) and will make it possible to identify new attractive poles (for example, the Covered Market area) and structures suitable for the creation of commercial parks.

The promotion of Gorizia's identity as an international city, connected to the Collio and Karst areas, associated with the promotion of the cultural, naturalistic, and food and wine offerings of the territory, are the other sectors to focus on.

Community Development

Alternatives will be evaluated based on the possibility of increasing the city's attractiveness to new populations, particularly the youth population. Attractiveness, in fact, is a strategic indicator of territorial competitiveness performance as it affects a territory's competitiveness, relationships between people and businesses, social and economic transactions, and the quality of the environment and life. The Municipality of Gorizia defines monitoring and evaluation indicators for actions to be implemented, based on the results of problem analysis, general objectives, and specific objectives.

The objectives are summarized as:

- Preventing the impoverishment of future generations;
- Improving the quality of life and equity among current generations;
- Being an active part in terms of ideas and stimuli in defining new projects in the territory;
- Building new alliances between local actors (social partnership).

Services for Citizens

Regarding the criterion of services for citizens, reference is made to improving their quality of life (improving the environmental, economic, and social dimensions). The Municipality of Gorizia proposes to evaluate alternatives based on their contribution to the following guidelines:

- Spreading knowledge in the community;
- Contributing to professional updating;
- Enhancing completed projects and those under execution;
- Promoting local typical products;
- Joining national networks (e.g.: National Coordination A21L) and European ones (e.g.: European Campaign for Sustainable Cities);
- Increasing visibility, commitment, and institutional positioning on Sustainable Development and Local Agenda 21 issues at provincial, regional, and national levels;
- Giving concreteness to its role as an agent of change and innovation at the local level.

11.1.3 Port System Authority Criteria

For the Port System Authority of the Eastern Adriatic, the criteria are:

- The centrality of the system and accessibility of inland ports
- Good service level
- Internalization of external costs.

The description of these criteria refers to the document: Port Regulatory Plan of Trieste, Resolution No. 36 of 27.04.2009 Agreement between the Municipality of Trieste and Port Authority, and the related SEA-EIA integrated procedure pursuant to Art. 6 paragraph 3 ter of Legislative Decree 152/2006.

Internalization of External Costs

This criterion outlines a vision for which the port system of Trieste becomes the center of relevant intermodal traffic flows, in connection with a broader system extended to Central Europe.

Alternatives will be evaluated in relation to the possibility of realizing a logistics platform capable of guaranteeing and expanding the centrality of the port system, where regional ports are understood as an essential part of the Upper Adriatic Port System and integrate with the system of inland port services based on truck stops, interports, and freight centers. In this context, the capacity to receive and forward goods, determined by port facilities, and the presence of both road and rail connections are the main indicators for the realization of a regional inland port platform.

Good Service Level

For the Port System Authority, achieving a good service level essentially concerns the movement of goods through:

- Increasing the efficiency of logistics services;
- Active and passive processing of goods, both with classic loading and unloading operations of intermodal units (empty and full containers from 20, 30, to 40 ft, tanks, swap bodies from 7.15 to 13.60 m) and conventional goods (timber, logs, pallets, big bags, coils, coils, cartons, bales, etc.);
- Strengthening road and rail connections, without excluding accessory services to goods and maintenance services.

Internalization of External Costs

Transport determines overall costs and benefits, understanding by the term "overall" the public and private, monetary and non-monetary, intentional and unintentional costs and benefits [1]. Its growing dimensions have made the interactions it has with the economic system and the social and natural environment increasingly complex. The negative (or positive) consequences of a transport activity, without the person who causes them (or benefits from them) bearing (or acquiring) a monetary compensation, are called externalities (generally defined in economics as a cost, or benefit, not paid by the subject who generates or receives it) [2].

The process through which it is possible to correct the allocative inefficiency deriving from the existence of externalities, which arises especially in the case of external diseconomies, is called internalization of externalities. The need to comply with it is claimed above all for environmental defense and for the need for a better quality of life, which is equivalent to optimizing logistics resources to the maximum, with efficiency and effectiveness.

For the Port System Authority, identifying options for internalization means:

- Promoting economic efficiency;
- Comparing alternative configurations of the transport system;
- Evaluating the equity of the distribution of costs and benefits.

Such operations stimulate productivity, which translates into economic growth as improved efficiency reduces costs. Therefore, according to European observations (High Level Group), priority must be given to economic and fiscal instruments (incentives and disincentives).

11.1.4 Friuli Venezia Giulia Region Criteria

For the FVG Region, the criteria are:

- Regional territorial competitiveness
- Decongestion of the heavy traffic system
- Infrastructure rebalancing
- The environmental issue.

The description of these criteria refers to documents within the Regional Infrastructure Plan for Transport, Goods Mobility, and Logistics, Regional Law of August 20, 2007, No. 23, Art. 3 ter and subsequent amendments.

Regional Territorial Competitiveness

This criterion aims to verify the contribution of alternatives in making Friuli-Venezia Giulia a competitive territory, capable of offering logistics services for the vast regional area consisting of Veneto, Carinthia, Slovenia, and Croatia, also by virtue of the implementation of new infrastructures provided by the Community programming of TEN networks (Priority Project No. 6) and the Adriatic-Baltic Corridor. A good level of regional territorial competitiveness can be achieved by focusing on efficient, punctual, and linear infrastructures, in such a way that Friuli-Venezia Giulia can become a hub of exchanges between Central-Eastern Europe, Northern Europe, the Mediterranean, and the Far East.

In evaluating the role of logistics on a national scale, the Friuli-Venezia Giulia Region refers to the orientation of the policy lines contained in the "New National Logistics Plan" (approved by the Council in December 2010), which identifies, among other things, the unitary logistics areas into which the national territory is divided, in order to make the logistics and infrastructure arrangement more effective.

In order to insert this "North East Logistics Platform" in a broad European context, international connections and those with contiguous regions, particularly Veneto and Slovenia, and certain poles of the Austrian territory, which are configured as essential hubs for the North–South directions of regional interest, must be strengthened.

In this perspective, it is appropriate to provide for a strengthening of existing infrastructures to allow increasingly accentuated forms of integration with the neighboring Republic of Slovenia, having as reference poles Trieste and Koper on one side, Gorizia and Nova Gorica on the other; in order to obtain a strengthening of the connections of areas adjacent to the border line, between which there already exists a strong osmosis.

In conclusion, the will of the Friuli-Venezia Giulia Region to aim at the creation of an integrated network system, the development of national, regional, and also international ports, and the strengthening of intermodality, both for goods and passengers, is therefore reaffirmed.

Decongestion of the Heavy Traffic System

This criterion aims to verify the contribution of alternatives in expanding the development of intermodal transport, aimed at rebalancing and decongesting heavy traffic. To this end, the Friuli-Venezia Giulia Region has shared and adopted the mission of the document "European transport policy at the 2010 horizon, the time of choice," relating to the so-called "White Paper on Transport" that the European Commission adopted in 2001 and which establishes new rules for the future of transport in Europe. It indicates the promotion and development of intermodal transport as elements of primary importance. The balance between modes of transport is considered one of the main EU objectives, and the incentive for intermodality is configured as a fundamental measure to develop alternatives to road transport.

This modal rebalancing is considered the preferred tool to cope with the growth of freight transport and to decongest the road system, in some important sections. The congestion of road traffic, in fact, constitutes a negative element in relation to two fundamental emergencies: environment and traffic safety.

The objective is to transfer increasingly consistent shares of freight traffic from road to rail and maritime transport. To this end, a framework is defined that provides for functional and multi-modal corridors dedicated primarily to goods, with the consequence of also lightening passenger traffic by road, considering that in the current situation there are no dedicated railway lines for either passengers or freight traffic.

In the medium and long term, the Region will be able to verify the opportunity, as well as the necessity, of building railway infrastructures dedicated exclusively to goods or passengers with a wide margin of acceptance of freight traffic in terms of capacity.

The performance of alternatives will be considered as a function of the regional strategy to reduce road traffic, enhancing rail traffic for connections with logistics centers and ports of the region, in order to make the goods mobility system more efficient.

Infrastructure Rebalancing

The uncontrolled growth, and not appropriately balanced, of transport infrastructures has produced considerable inconsistencies in the system itself and has contributed decisively to producing the current level of environmental risk, with direct and indirect consequences on citizens' lives. Therefore, it is necessary, in order to ensure a rebalancing between the different modes of transport, to eliminate the bottlenecks that make the system inefficient and simultaneously place citizens and the environment at the center of transport policy. Modal rebalancing is both an objective of transport policy and of the Plan.

It is therefore appropriate to carry out an infrastructural rebalancing of the territory, both at a regional level and at a sub-regional level, with a view to social cohesion without neglecting local economic needs. Overcoming the infrastructural gap for sub-regional areas of strong productive value means, for the Region's economic system, planning the endowment of road infrastructures for connection to the main road arches from/to metropolitan areas and other regions.

Furthermore, within the framework of issues related to passenger transport, the increase in demand induced by the enlargement of Europe has increased the request for accessibility of both metropolitan and peripheral areas. The numerical values of these flows and their growing trend over time impose radical and structured choices that are defined in a higher context.

To respond to the request for accessibility of metropolitan areas, which by their nature are articulated and complex, it is necessary to make the transport system, both in the rubber mode and in that of the railway, more efficient in urban and metropolitan areas, faster, safer, and that guarantees a substantial reduction in pollution over time.

In the Plan of the transport system of the Friuli-Venezia Giulia Region, the necessary interventions of an exquisitely infrastructural nature are identified. They mainly provide for completions, adjustments, and some operational recoveries. Numerous interventions capable of improving interconnections to be implemented in terms of services, particularly as regards their rationalization and strengthening, are also to be considered indispensable.

The Environmental Issue

The environmental issue has assumed a predominant role, as it can no longer be postponed, given the enormous contribution to environmental pollution of road transport. Furthermore, this criterion can become the driving force of the Plan in its road section, given its strong relationship to the environmental component. The environment and its quality are recognized as scarce resources to be protected, so much so that they are at the center of the most important current international agreements prepared in compliance with the principle of sustainability. The approach consists of identifying a correct strategic and economic dimension that the transport sector must play within the framework of sustainable policies for economic, social, and environmental development of the regional territory.

Conditions for sustainable development are, in fact, the guarantee of a high level of environmental protection and their integration with strictly productive-economic considerations, given that, to guarantee the sustainability of choices, it is necessary to minimize their impacts on the environment and suggest appropriate environmental and ecological mitigation measures.

The new interventions to protect the environmental sustainability of the territory imply the need to contain the consumption of petroleum products, linked to that of reducing CO_2 emissions, attributing growing opportunities to Mediterranean ports, and within it to the Adriatic one.

This criterion regulates in particular the transport system in order to ensure the fundamental right of citizens to mobility, contribute to environmental protection, promote balanced economic and social development, favoring in particular the integration of different means of transport and simultaneously pursuing rationalization and effectiveness of expenditure.

11.2 Technique for Determining Weights: Simplified Simos Method

At this stage, criteria weighting is carried out, a crucial step in multi-criteria decision support models (MCA). The weight assigned to each criterion facilitates comparisons between criteria and explicitly reflects the decision-maker's preferences [3].

As outlined in Chap. 7, the Simplified Simos Method was chosen for this analysis—a simplified calculation procedure proposed by Crisman et al. [4].

11.2.1 Application

1. During one of the working sessions, decision-makers were asked to rank the identified criteria from least to most important, allowing for ties where necessary. Additionally, they could insert one or more "blank cards" in the ranking to represent differences in importance between adjacent criteria.
2. Each stakeholder established their own ranking as follows (Table 11.1).

   At this juncture, the weighting procedure for criteria is elicited for each stakeholder as described in Chap. 7. The following pages illustrate the calculation methodology using SDAG as an exemplary stakeholder, while the complete results of the weight elicitation process for all stakeholders can be found in the subsequent section.
3. The evaluator first assigns a rank (r) to each level in the table, designating a value of 1 to the least important criterion and then increasing ordinally, taking into account levels with blank cards. Subsequently, the evaluator applies a value (vr) associated with the rank of each criterion, excluding blank cards. In this method, the values associated with each step (vr) represent an initial, non-normalized judgment of the relative importance of the various criteria under consideration (Table 11.2).
4. The importance ratio (z) between the most important criterion and the least important criterion is defined, in this case z = 7.
5. The weight (w_r) is calculated using linear interpolation to maintain a ratio between the weight associated with the most important criterion and the least important criterion equal to z:

$$w_r = \frac{v_r(z-1)}{v_{rmax} - v_{rmin}} + \frac{v_{rmax} - zv_{rmin}}{v_{rmax} - v_{rmin}}$$

Table 11.1 Ranking of criteria for stakeholders

Stakeholder	Criteria from least important to most important
SDAG	Economic development of the territory Increase in activities performed by SDAG Economic incentives *Blank card* Establishment of new clients in SDAG Overcoming marginality Budget balance
Municipality of Gorizia	Services for citizens *Blank card* Community development Economic and entrepreneurial development
Port system authority	Internalization of external costs *Blank card* *Blank card* Good service level *Blank card* The centrality of the system and accessibility of inland ports
Friuli Venezia Giulia Region	The environmental issue *Blank card* Decongestion of the heavy traffic system Infrastructure rebalancing *Blank card* Regional territorial competitiveness

Table 11.2 Illustration of step 3 of the simplified Simos method

Stakeholder	Criteria from least important to most important	Rank	Rank's value
		r	v_r
SDAG	Economic development of the territory	1	1
	Increase in activities performed by SDAG	2	2
	Economic incentives	3	3
	Blank card	4	–
	Establishment of new clients in SDAG	5	5
	Overcoming marginality	6	6
	Budget balance	7	7

where w_r is the weight of the criterion; z is the ratio between the most important and least important criteria; v_r is the value associated with the rank (Table 11.3).

6. The rank weight (w_t) is calculated by multiplying the weight of criteria at a given rank by the number of criteria at that same step. In this case, each step contains at most one criterion, therefore $w_t = w_r$. The values obtained are then summed, resulting in $\sum w_t = 24$.

Table 11.3 Illustration of step 5 of the simplified Simos method

Stakeholder	Criteria from least important to most important	Rank	Rank's value	Weight
		r	v_r	w_r
SDAG	Economic development of the territory	1	1	1.0
	Increase in activities performed by SDAG	2	2	2.0
	Economic incentives	3	3	3.0
	Blank card	4	–	
	Establishment of new clients in SDAG	5	5	5.0
	Overcoming marginality	6	6	6.0
	Budget balance	7	7	7.0

Table 11.4 Illustration of step 7 of the simplified Simos method

Stakeholder	Criteria from least important to most important	Rank	Rank's value	Criteria's	Normalized weight
		r	vr	w_r	w_r^n
SDAG	Economic development of the territory	1	1	1.0	0.04
	Increase in activities performed by SDAG	2	2	2.0	0.08
	Economic incentives	3	3	3.0	0.13
	Blank card	4	–	–	–
	Establishment of new clients in SDAG	5	5	5.0	0.21
	Overcoming marginality	6	6	6.0	0.25
	Budget balance	7	7	7.0	0.29

7. Finally, the normalized weight (w_r^n) is calculated by normalizing each w_r by $\sum w_t$ (Table 11.4).

11.2.2 Results

The following sections present the results of the criteria weight elicitation procedure for each stakeholder (Table 11.5).

Table 11.5 Results of simplified Simos method

Stakeholder	Criteria from least important to most important	Rank	Rank's value	Criteria's weight	Normalized weight
		r	vr	w_r	w_r^n
SDAG	Economic development of the territory	1	1	1.0	0.04
	Increase in activities performed by SDAG	2	2	2.0	0.08
	Economic incentives	3	3	3.0	0.13
	Blank card	4	–	–	–
	Establishment of new clients in SDAG	5	5	5.0	0.21
	Overcoming marginality	6	6	6.0	0.25
	Budget balance	7	7	7.0	0.29
Municipality of Gorizia	Services for citizens	1	1	1.0	0.1
	Blank card	2	–	–	–
	Community development	3	3	3.0	0.4
	Economic and entrepreneurial development	4	4	4.0	0.5
Port system authority	Internalization of external costs	1	1	1	0.09
	Blank card	2	–	–	–
	Blank card	3	–	–	–
	Good service level	4	4	4	0.36
	Blank card	5	–	–	–
	Centrality of the system and accessibility of inland ports	6	6	6	0.55
Friuli Venezia Giulia Region	The environmental issue	1	1	1	0.07
	Blank card	2	–	–	–

(continued)

Table 11.5 (continued)

Stakeholder	Criteria from least important to most important	Rank	Rank's value	Criteria's weight	Normalized weight
		r	vr	w_r	w_r^n
	Decongestion of the heavy traffic system	3	3	3	0.21
	Infrastructure rebalancing	4	4	4	0.29
	Blank card	5	–	–	–
	Regional territorial competitiveness	6	6	6	0.43

11.2.2.1 SDAG

For SDAG, the weight calculation procedure considers six criteria with a relative importance ratio (z) of 7 between the most and least important criteria. In this case, $\sum w_t$ equals 24.

The analysis reveals that SDAG assigns the highest importance to budget balance and the lowest to territorial economic development, with the establishment of new activities positioned at an intermediate level. It becomes evident that achieving budget balance constrains other criteria, and that SDAG aims to overcome marginality (the second most important criterion) through attracting new clients rather than making larger investments from operating revenues, in order not to compromise budget equilibrium (Fig. 11.1).

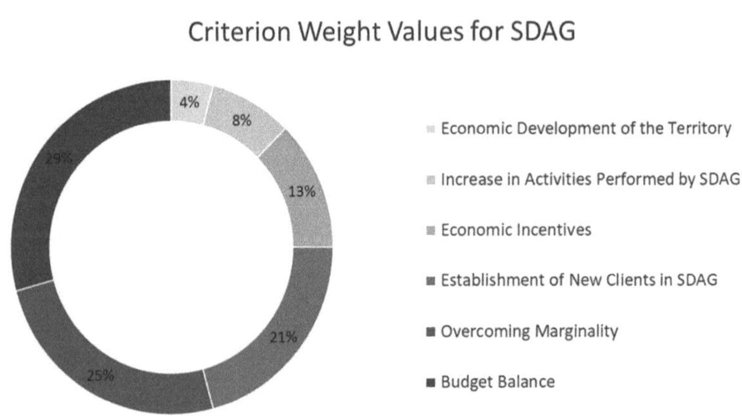

Fig. 11.1 Area chart of criterion weight values for SDAG

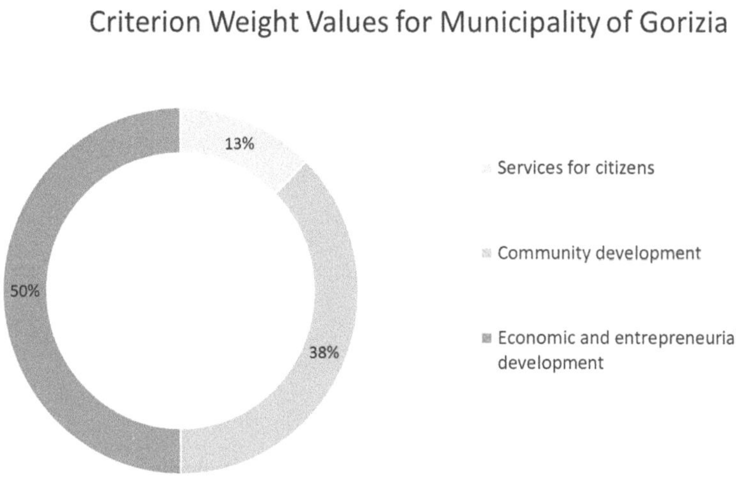

Fig. 11.2 Area chart of criterion weight values for Municipality of Gorizia

11.2.2.2 Municipality of Gorizia

For the Municipality of Gorizia, the weight calculation procedure examines three criteria with a relative importance ratio (z) of 4 between the most and least important criteria. In this case, $\sum w_t$ equals 8.

The analysis demonstrates that the Municipality of Gorizia attributes greater value to criteria related to economic development and less importance to citizen services, focusing more on community development. The Municipality appears to believe that the community can benefit from economic development through increased overall welfare (Fig. 11.2).

11.2.2.3 Port System Authority

For the Port System Authority, the weight calculation procedure considers three criteria with a relative importance ratio (z) of 6 between the most and least important criteria. In this case, $\sum w_t$ equals 11.

The Port System Authority gives maximum priority to system centrality, which it aims to achieve through a high level of service, while considering the internalization of external costs to be of minor significance (Fig. 11.3).

11.2.2.4 Friuli Venezia Giulia Region

For the Friuli Venezia Giulia Region, the weight calculation procedure evaluates four criteria with a relative importance ratio (z) of 6 between the most and least important criteria. In this case, $\sum$wt equals 14.

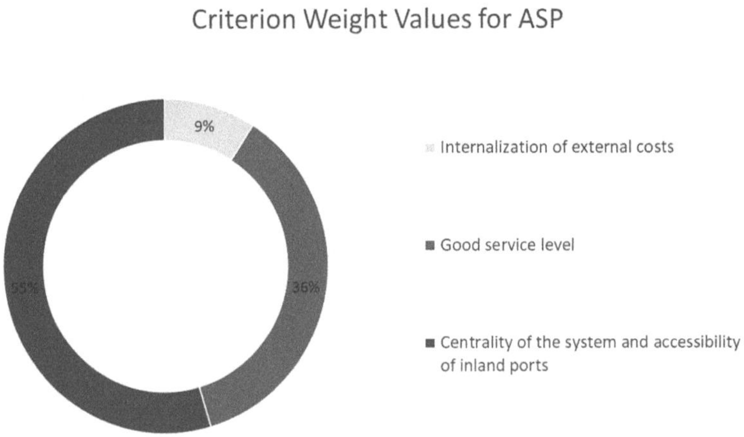

Fig. 11.3 Area chart of criterion weight values for Port System Authority

The Friuli Venezia Giulia Region places territorial competition at the top of its criteria weight ranking and environmental issues at the bottom. The Region appears to address environmental concerns through the decongestion of heavy traffic, while focusing on territorial competitiveness through infrastructural rebalancing, which ranks second in its priority hierarchy (Fig. 11.4).

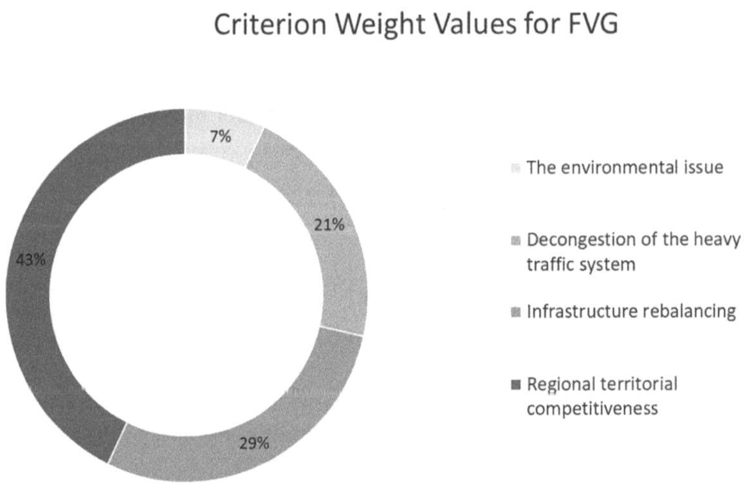

Fig. 11.4 Area chart of criterion weight values for Friuli Venezia Giulia Region

11.3 Identification of Temporal Framework

The specification of the temporal framework (short, medium, and long term) is modeled on the planning adopted by stakeholders, as the performance of an alternative depends not only on its inherent quality but also on the quality of internal and external information—both contingent and contextual—that the decision-making process makes available [5].

For the short term, a three-year timeframe has been established because the Port System Authority has developed a 2021–2023 Performance Plan. For the medium term, a seven-year interval has been selected, as SDAG prepares a seven-year programmatic document. Finally, for the long term, reference has been made to the Green Book, which considers a 10-year time horizon as an appropriate working hypothesis for many infrastructure interventions [6]. Local and regional planning does not expire until the implementation of a subsequent plan, but it is reasonable to establish a 10-year time limit beyond which the forecast will no longer be relevant [6].

Consequently, the temporal framework is structured as follows.

In detail, the time horizon starts from moment $t = 0$ in the year 2021 and continues as highlighted in the Table 11.6.

Table 11.6 Time horizon and effects considered for this evaluation

Period	Number of years	Years	Effects considered
Present	1	2021	At the present moment, only the performance of the zero alternative is considered for all alternatives, as the others have not yet been implemented
Short term	3	2022–2024	The effects of the short term are considered starting from the third year (2024); for the period 2022–2023, it is reasonable to consider only the performance of the zero alternative
Medium term	7	2025–2031	The performance of the medium term takes effect only starting from the tenth year (2031); for the period 2025–2030, it is reasonable to consider the performance of alternatives for the short term
Long term	10	2032–2041	The performance of the long term takes effect only starting from the twentieth year (2041); for the period 2032–2040, it is reasonable to consider the performance of alternatives for the medium term
Long term effects	10	2042–2051	It is reasonable to consider the performance of the long term valid for an interval of 10 years, ending in 2051

11.4 Evaluation Matrices

Building upon the evaluation criteria and their respective weights, the alternatives have been systematically analyzed and evaluated. The following Tables 11.7, 11.8 and 11.9 present the performance matrices, which assign scores to each alternative on a scale ranging from 0 to 5.

As mentioned in Chap. 7, the Simple Additive Weighting (SAW) linear additive method has been employed at this stage. For a more comprehensive understanding, please refer to Chap. 3 Simple Additive Weighting method.

The subsequent sections will present the results of this procedure for each stakeholder across various time periods. As previously noted, only the performance of alternative 0 is considered for the present period.

11.4.1 SDAG Preference

For SDAG, the GoFoodLog alternative emerges as the preferred option in the short term and maintains this position throughout both the medium and long term, as it consistently achieves the highest values across all criteria. The zero alternative consistently ranks last, marginally outperformed by the intermodal terminal. As time

Table 11.7 Short term performance for SDAG

Short-term performance for SDAG						
Criteria	Criterion weight (%)	Alternatives				
		Alternative 0	GoFooodLog	Safe and secure parking	ZLS	Intermodal terminal
Budget balance	29	1.46	1.46	1.46	1.46	1.46
Economic incentives	13	0.00	0.63	0.63	0.63	0.63
Overcoming marginality	25	0.00	1.00	1.25	0.75	0.00
Increase in activities performed by SDAG	8	0.00	0.25	0.08	0.00	0.00
Establishment of new clients in SDAG	21	0.00	0.83	0.21	0.42	0.00
Economic development of the territory	4	0.00	0.08	0.04	0.08	0.00
Total		1.46	4.25	3.67	3.33	2.08

Table 11.8 Medium term performance for SDAG

Medium-term performance for SDAG						
Criteria	Criterion weight (%)	Alternatives				
		Alternative 0	GoFooodLog	Safe and secure parking	ZLS	Intermodal terminal
Budget balance	29	1.46	1.46	1.46	1.46	1.46
Economic incentives	13	0.00	0.63	0.63	0.63	0.63
Overcoming marginality	25	0.00	1.00	0.75	0.75	0.50
Increase in activities performed by SDAG	8	0.00	0.33	0.08	0.00	0.08
Establishment of new clients in SDAG	21	0.00	1.04	0.63	0.83	0.21
Economic development of the territory	4	0.00	0.13	0.04	0.13	0.04
Total		1.46	4.58	3.58	3.79	2.92

Table 11.9 Long term performance for SDAG

Long-term performance for SDAG						
Criteria	Criterion weight (%)	Alternatives				
		Alternative 0	GoFooodLog	Safe and secure parking	ZLS	Intermodal terminal
Budget balance	29	1.46	1.46	1.46	1.46	1.46
Economic incentives	13	0.00	0.63	0.63	0.63	0.63
Overcoming marginality	25	0.00	0.75	0.75	0.75	0.75
Increase in activities performed by SDAG	8	0.00	0.25	0.08	0.00	0.17
Establishment of new clients in SDAG	21	0.00	0.42	0.63	0.63	0.42
Economic development of the territory	4	0.00	0.13	0.04	0.13	0.08
Total		1.46	3.63	3.58	3.58	3.50

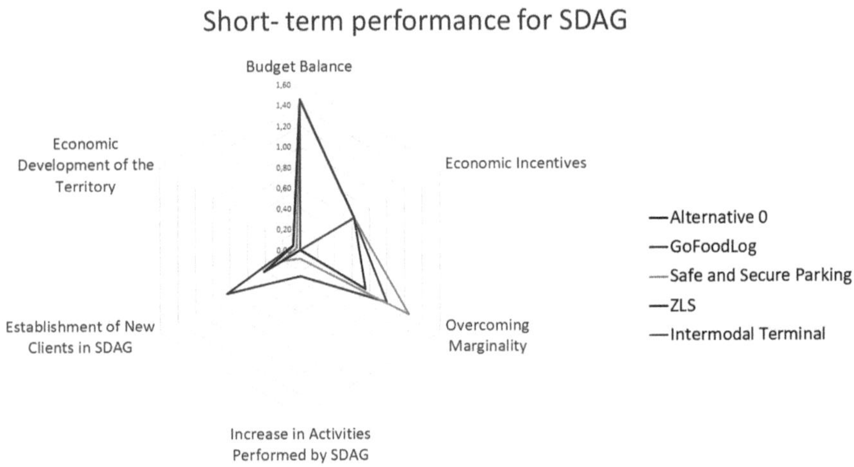

Fig. 11.5 Radar graph of short-term performance for SDAG

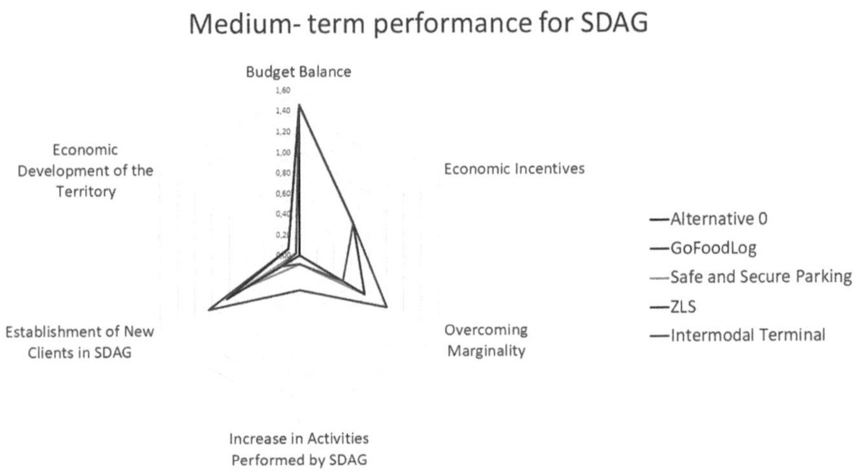

Fig. 11.6 Radar graph of medium-term performance for SDAG

progresses, the intermodal terminal gains in importance, with results converging toward a uniform value in the long term (Figs. 11.5, 11.6, 11.7, 11.8).

11.4.2 *Municipality of Gorizia Preference*

For the Municipality of Gorizia, the GoFoodLog alternative maintains a constant score of 3.13 across all three periods under consideration. While this represents the

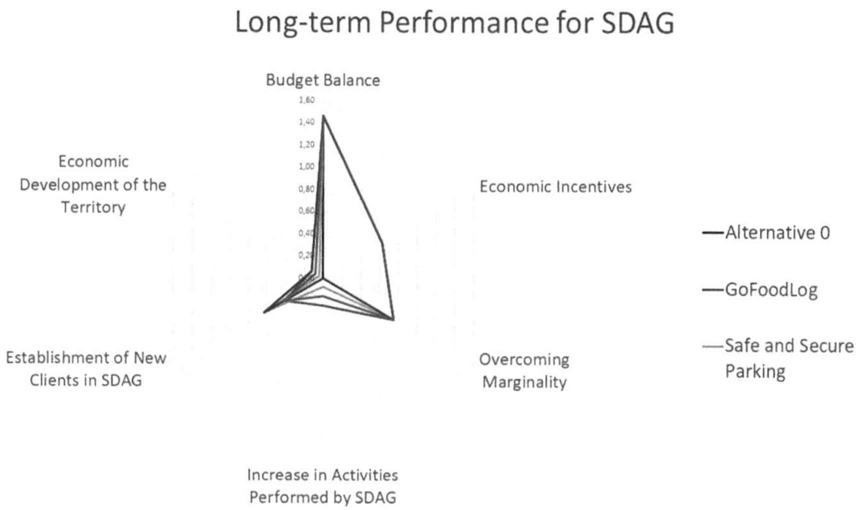

Fig. 11.7 Radar graph of long-term performance for SDAG

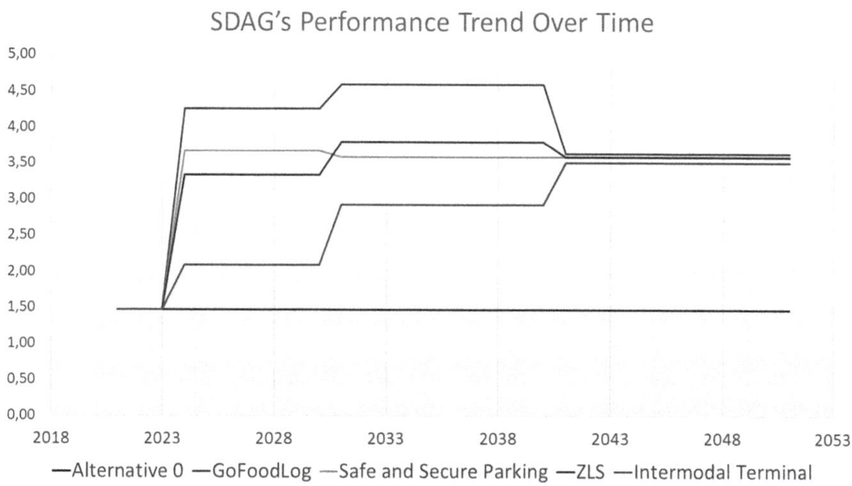

Fig. 11.8 Performance trend over time for SDAG

highest score in the short term, it is subsequently surpassed by the ZLS alternative in both the medium and long term. The Safe and Secure Park alternative consistently occupies the penultimate position, while the zero alternative, with null performance values, remains in last place (Figs. 11.9, 11.10, 11.11, 11.12 and Tables 11.10, 11.11, 11.12).

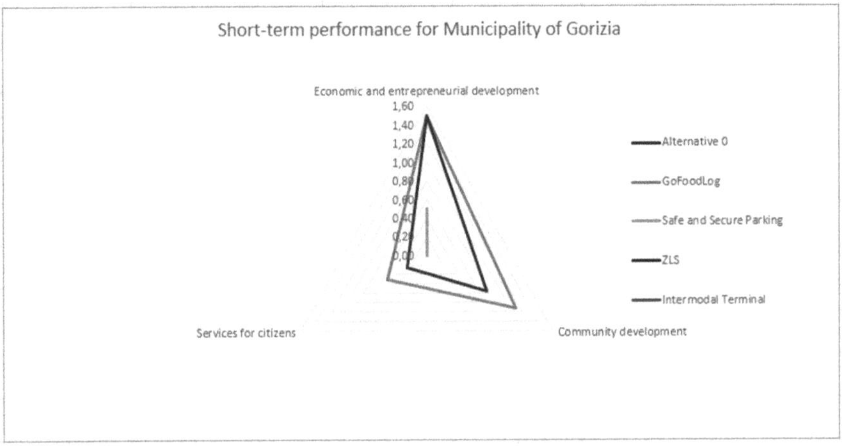

Fig. 11.9 Radar graph of short-term performance for Municipality of Gorizia

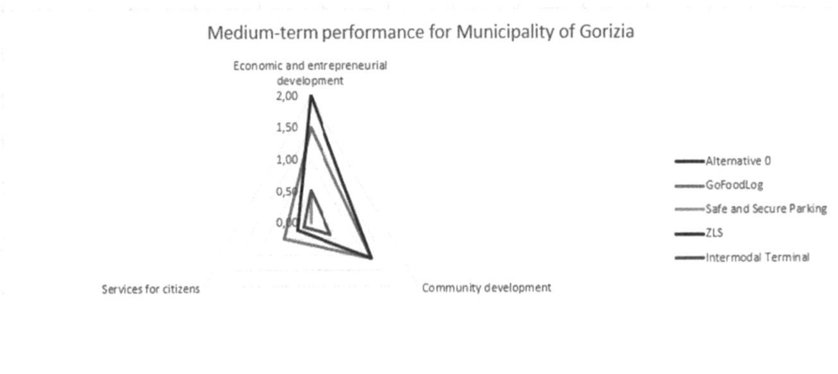

Fig. 11.10 Radar graph of medium-term performance for Municipality of Gorizia

11.4.3 Port System Authority Preferences

For the Port System Authority, the GoFoodLog alternative consistently ranks first across all periods under consideration, despite experiencing a slight decline in the final period. The ZLS alternative maintains a steady second-place position and strengthens in the medium and long term, though never overtaking GoFoodLog. The Safe and Secure Park remains consistently the least appealing option, though still outperforming the zero alternative, which receives no points (Figs. 11.13, 11.14, 11.15, 11.16 and Tables 11.13, 11.14, 11.15).

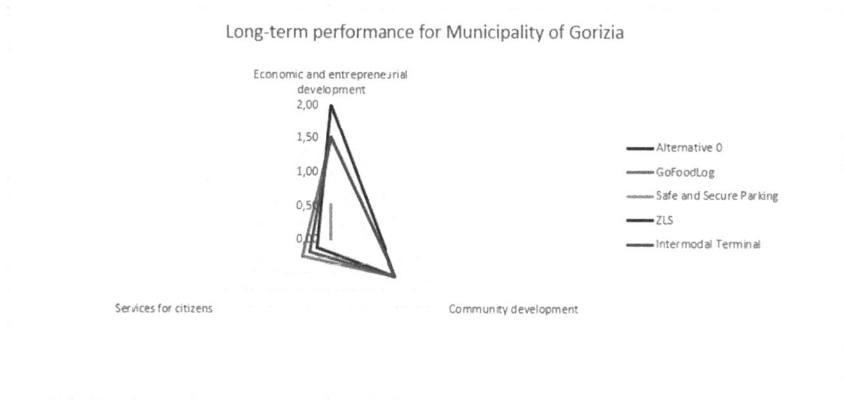

Fig. 11.11 Radar graph of long-term performance for Municipality of Gorizia

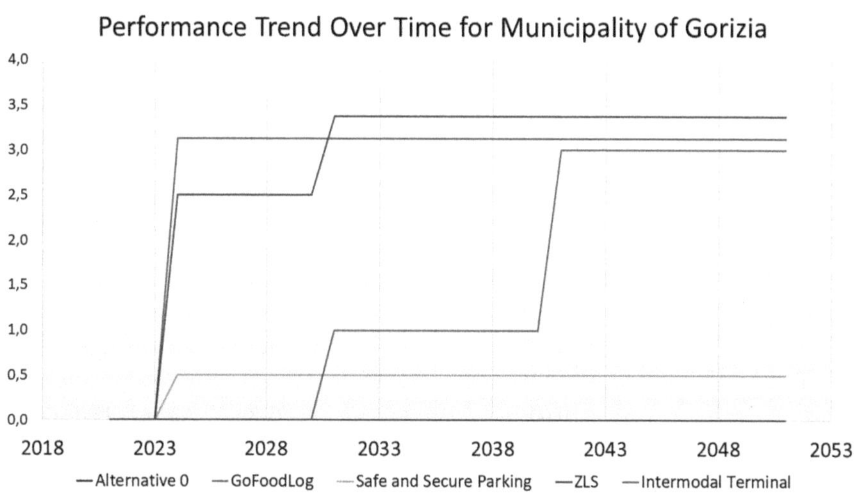

Fig. 11.12 Performance trend over time for Municipality of Gorizia

11.4.4 Friuli Venezia Giulia Region Preferences

For the Friuli Venezia Giulia Region, the most significant alternative is ZLS, though it declines in the long term, with GoFoodLog occupying second place. In the short term, Safe and Secure Park emerges as the least interesting option, considering that the zero alternative never receives any points. The intermodal terminal decreases in importance in the medium term before recovering in the long term. In the long term, Safe and Secure Park ranks in the penultimate position, surpassed by the intermodal terminal (Figs. 11.17, 11.18, 11.19, 11.20 and Tables 11.16, 11.17, 11.18).

Table 11.10 Short term performance for Municipality of Gorizia

Short-term performance for Municipality of Gorizia						
Criteria	Criterion weight (%)	Alternatives				
		Alternative 0	GoFooodLog	Safe and secure parking	ZLS	Intermodal terminal
Economic and entrepreneurial development	50	0.00	1.50	0.50	1.50	0.00
Community development	38	0.00	1.13	0.00	0.75	0.00
Services for citizens	13	0.00	0.50	0.00	0.25	0.00
Total		0.00	3.13	0.50	2.50	0.00

Table 11.11 Medium term performance for Municipality of Gorizia

Medium-term performance for Municipality of Gorizia						
Criteria	Criterion weight (%)	Alternatives				
		Alternative 0	GoFooodLog	Safe and secure parking	ZLS	Intermodal terminal
Economic and entrepreneurial development	50	0.00	1.50	0.50	2.00	0.50
Community development	38	0.00	1.13	0.00	1.13	0.38
Services for citizens	13	0.00	0.50	0.00	0.25	0.13
Total		0.00	3.13	0.50	3.38	1.00

Table 11.12 Long term performance for Municipality of Gorizia

Long-term performance for Municipality of Gorizia						
Criteria	Criterion weight (%)	Alternatives				
		Alternative 0	GoFooodLog	Safe and secure parking	ZLS	Intermodal terminal
Economic and entrepreneurial development	50	0.00	1.50	0.50	2.00	1.50
Community development	38	0.00	1.13	0.00	1.13	1.13
Services for citizens	13	0.00	0.50	0.00	0.25	0.38
Total		0.00	3.13	0.50	3.38	3.00

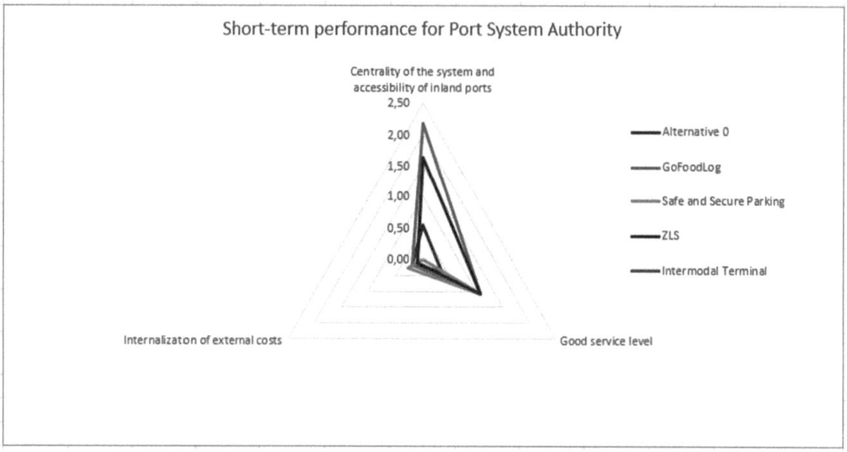

Fig. 11.13 Radar graph of short-term performance for Port System Authority

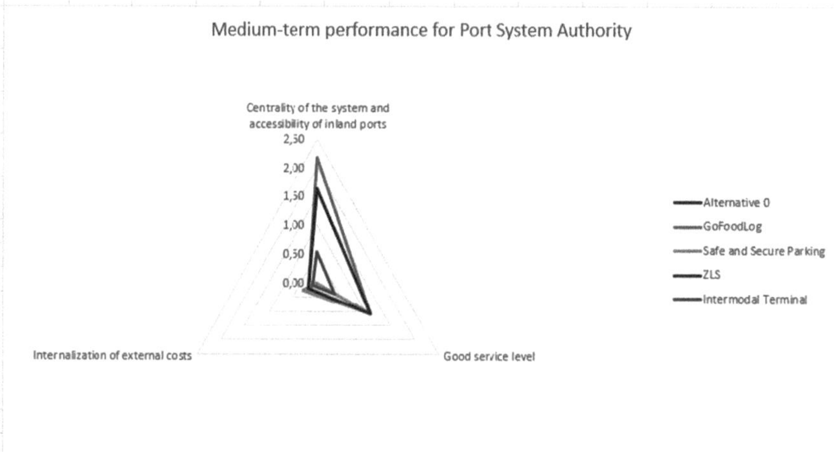

Fig. 11.14 Radar graph of medium-term performance for Port System Authority

11.4.5 Results and General Observations

A general observation reveals that all alternatives experience a significant decline in the long term, with the intermodal terminal being the only option that returns to values comparable with previous periods. The initial results are organized according to a temporal framework (Table 11.19).

For SDAG, the most favourable alternative is GoFoodLog, while the least preferred is the zero alternative. Similarly, for the Municipality of Gorizia,

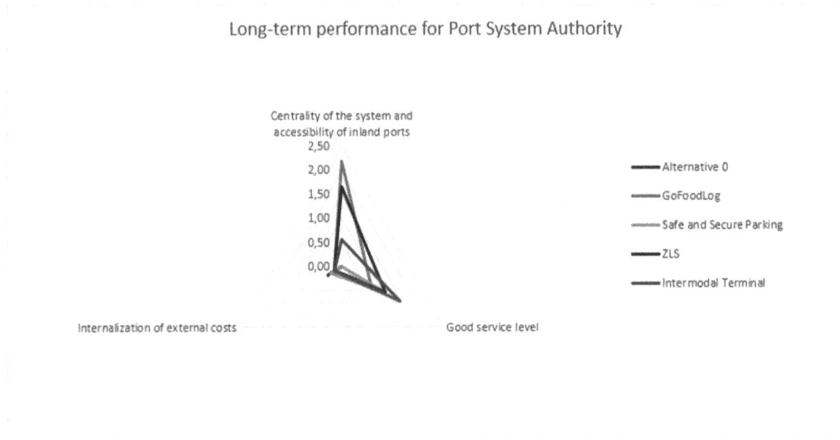

Fig. 11.15 Radar graph of long-term performance for Port System Authority

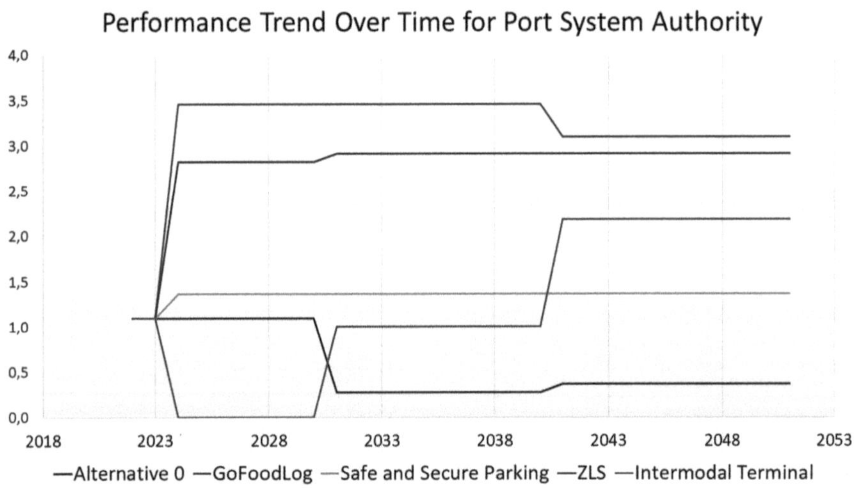

Fig. 11.16 Performance trend over time for Port System Authority

GoFoodLog emerges as the most viable option, whereas both the zero and intermodal alternatives are the least favourable. The Port System Authority also identifies GoFoodLog as the most promising alternative, while the intermodal terminal ranks lowest in preference. In contrast, the Friuli Venezia Giulia Region considers the Simplified Logistics Zone (ZLS) the most advantageous option, with the intermodal terminal being the least preferred (Table 11.20).

When considering a broader temporal perspective (medium term), SDAG continues to favor GoFoodLog, with the zero alternative remaining the least appealing. The Municipality of Gorizia maintains its preference for GoFoodLog

Table 11.13 Short term performance for Port System Authority

Short-term performance for Port System Authority						
Criteria	Criterion weight (%)	Alternatives				
		Alternative 0	GoFooodLog	Safe and secure parking	ZLS	Intermodal terminal
Centrality of the system and accessibility of inland ports	55	0.55	2.18	0.00	1.64	0.00
Good service level	36	0.36	1.09	1.09	1.09	0.00
Internalization of external costs	9	0.18	0.18	0.27	0.09	0.00
Total		1.09	3.45	1.36	2.82	0.00

Table 11.14 Medium term performance for port system authority

Medium-term performance for Port System Authority						
Criteria	Criterion weight (%)	Alternatives				
		Alternative 0	GoFooodLog	Safe and secure parking	ZLS	Intermodal terminal
Centrality of the system and accessibility of inland ports	55	0.00	2.18	0.00	1.64	0.55
Good service level	36	0.00	1.09	1.09	1.09	0.36
Internalization of external costs	9	0.27	0.18	0.27	0.18	0.09
Total		0.27	3.45	1.36	2.91	1.00

while deeming the zero alternative the least viable. Similarly, the Port System Authority prioritizes GoFoodLog, though both the zero alternative and the intermodal terminal are regarded as the least favorable. The Friuli Venezia Giulia Region remains consistent in its preference for ZLS, ranking the zero alternative lowest (Table 11.21).

When considering a broader temporal perspective (medium term), SDAG continues to favor GoFoodLog, with the zero alternative remaining the least appealing. The Municipality of Gorizia maintains its preference for GoFoodLog while deeming the zero alternative the least viable. Similarly, the Port System

Table 11.15 Long term performance for Port System Authority

Long-term performance for Port System Authority						
Criteria	Criterion weight (%)	Alternatives				
		Alternative 0	GoFooodLog	Safe and secure parking	ZLS	Intermodal terminal
Centrality of the system and accessibility of inland ports	55	0.00	2.18	0.00	1.64	0.55
Good service level	36	0.00	0.73	1.09	1.09	1.45
Internalization of external costs	9	0.36	0.18	0.27	0.18	0.18
Total		0.36	3.09	1.36	2.91	2.18

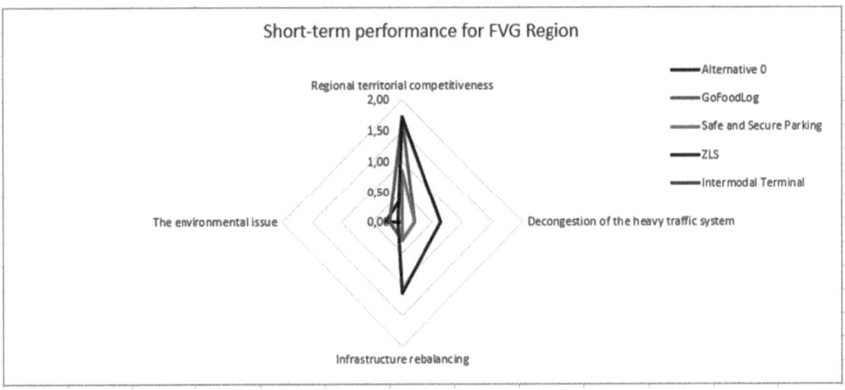

Fig. 11.17 Radar graph of short-term performance for FVG Region

Authority prioritizes GoFoodLog, though both the zero alternative and the intermodal terminal are regarded as the least favorable. The Friuli Venezia Giulia Region remains consistent in its preference for ZLS, ranking the zero alternative lowest.

11.4.6 Overall Considerations

To summarize, SDAG identifies the zero alternative as the least effective, whereas the remaining options, despite exhibiting varying performance levels in the short and medium term, tend to converge toward a singular value in the long term.

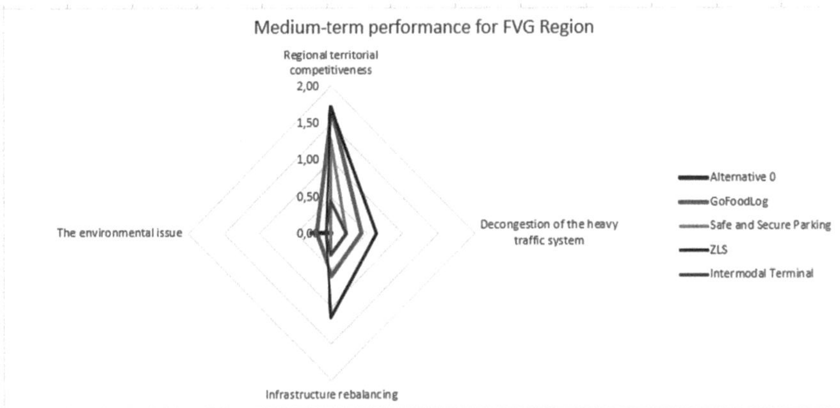

Fig. 11.18 Radar graph of medium-term performance for FVG Region

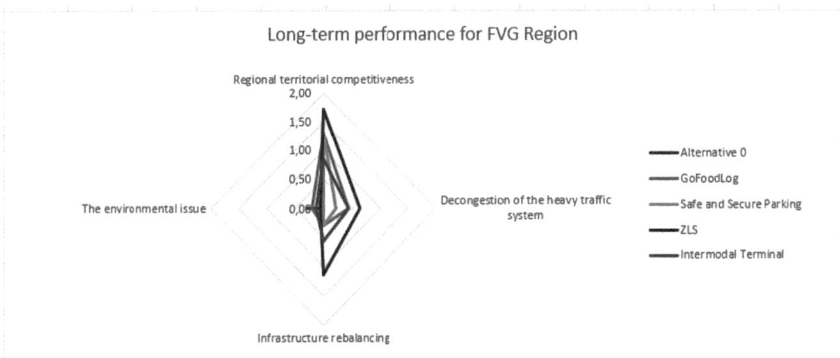

Fig. 11.19 Radar graph of long-term performance for FVG Region

For the Municipality of Gorizia, the zero alternative proves entirely ineffective. The most significant divergence from SDAG's preferences lies in the evaluation of the Safe and Secure Parking alternative, which offers no tangible benefits to the region but only to the entity managing the facilities. Ultimately, the alternative that best aligns with the Municipality's criteria is the Simplified Logistics Zone (ZLS).

The Port System Authority, on the other hand, regards GoFoodLog as the most advantageous option, despite its performance not demonstrating consistent growth over time. Notably, in the long run, there is a discernible decline in its performance.

Similar to the Municipality of Gorizia, the Friuli Venezia Giulia Region identifies the Strengthened Simplified Logistics Zone as the most preferable alternative, given its superior and stable performance over time.

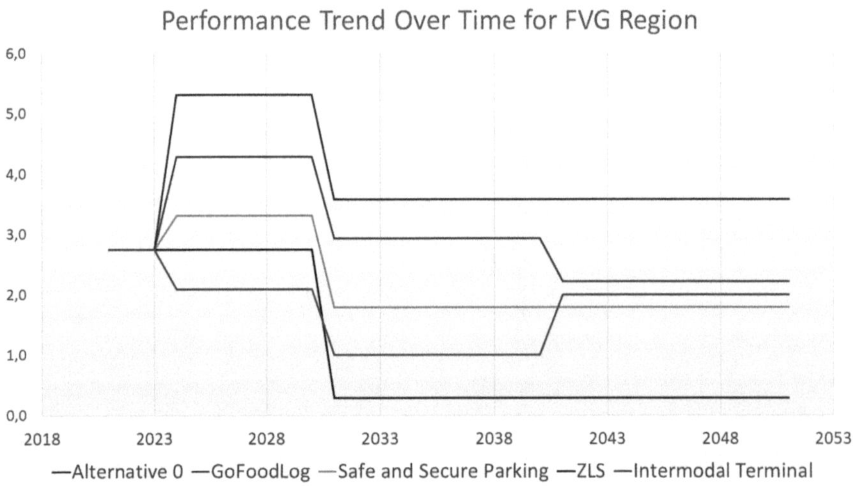

Fig. 11.20 Performance trend over time for FVG Region

Table 11.16 Short term performance for FVG Region

Short-term performance for FVG Region						
Criteria	Criterion weight (%)	Alternatives				
		Alternative 0	GoFooodLog	Safe and secure parking	ZLS	Intermodal terminal
Regional territorial competitiveness	43	0.43	1.71	0.86	1.71	0.00
Decongestion of the heavy traffic system	21	0.00	0.21	0.21	0.64	0.00
Infrastructure rebalancing	29	0.00	0.29	0.29	1.14	0.00
The environmental issue	7	0.29	0.21	0.00	0.07	0.00
Total		0.71	2.43	1.36	3.57	0.00

11.5 Determining Temporal Value

At this stage, it is necessary to adopt a descriptive model in order to address various aspects related to regional competitiveness, territorial and community development, and the quality of public services, while also considering environmental impacts. These dimensions cannot be fully captured or classified solely through

Table 11.17 Medium term performance for FVG Region

Medium-term performance for FVG Region						
Criteria	Criterion weight (%)	Alternatives				
		Alternative 0	GoFooodLog	Safe and secure parking	ZLS	Intermodal terminal
Regional territorial competitiveness	43	0.00	1.71	1.29	1.71	0.43
Decongestion of the heavy traffic system	21	0.00	0.43	0.21	0.64	0.21
Infrastructure rebalancing	29	0.00	0.57	0.29	1.14	0.29
The environmental issue	7	0.29	0.21	0.00	0.07	0.07
Total		0.29	2.93	1.79	3.57	1.00

Table 11.18 Long term performance for FVG Region

Long-term performance for FVG Region						
Criteria	Criterion weight (%)	Alternatives				
		Alternative 0	GoFooodLog	Safe and secure parking	ZLS	Intermodal terminal
Regional territorial competitiveness	43	0.00	1.29	1.29	1.71	0.86
Decongestion of the heavy traffic system	21	0.00	0.43	0.21	0.64	0.43
Infrastructure rebalancing	29	0.00	0.29	0.29	1.14	0.57
The environmental issue	7	0.29	0.21	0.00	0.07	0.14
Total		0.29	2.21	1.79	3.57	2.00

an nondimensional model, which is effective primarily when applied to monetary values.

Table 11.19 Alternative's performance in short-term

Short-term performance					
Stakeholder	Alternatives				
	Alternative 0	GoFooodLog	Safe and secure parking	ZLS	Intermodal terminal
SDAG	1.46	4.25	3.67	3.33	2.08
GO	0.00	3.13	0.50	2.50	0.00
PAS	1.09	3.45	1.36	2.82	0.00
FVG	0.71	2.43	1.36	3.57	0.00

Table 11.20 Alternative's performance in medium-term

Medium-term performance					
Stakeholder	Alternatives				
	Alternative 0	GoFooodLog	Safe and secure parking	ZLS	Intermodal terminal
SDAG	1.46	4.58	3.58	3.79	2.92
GO	0.00	3.13	0.50	3.38	1.00
PAS	0.27	3.45	1.36	2.91	1.00
FVG	0.29	2.93	1.79	3.57	1.00

Table 11.21 Alternative's performance in long-term

Long-term performance					
Stakeholder	Alternatives				
	Alternative 0	GoFooodLog	Safe and secure parking	ZLS	Intermodal terminal
SDAG	1.46	3.63	3.58	3.58	3.50
GO	0.00	3.13	0.50	3.38	3.00
PAS	0.36	3.09	1.36	2.91	2.18
FVG	0.29	2.21	1.79	3.57	2.00

To achieve a comprehensive evaluation, it is appropriate to use the intertemporal weighting coefficient (τ_r) as outlined in Chap. 8. This coefficient enables the comparison of highly heterogeneous performance outcomes occurring over different time periods and across varying temporal horizons. The coefficient (τ_r) is applied to the performance of individual criteria on an annual basis, variations depending on whether the criteria pertain to material aspects or to social and environmental factors.

Using the formulas detailed in Chap. 8, the intertemporal evaluation coefficient is calculated. The following Table 11.22 Temporal Horizon and Intertemporal Weighting Coefficients presents the year-by-year values. Specifically, the coefficient

$\tau_r{}^e$ is applied to material criteria, whereas the coefficient $\tau_r{}^h$ is used for criteria related to well-being.

Table 11.22 Temporal horizon and intertemporal weighting coefficients

t	Year	τ_r^e	τ_r^h
0	2021	1.000	1.000
1	2022	0.966	0.985
2	2023	0.934	0.971
3	2024	0.902	0.957
4	2025	0.871	0.943
5	2026	0.842	0.930
6	2027	0.814	0.917
7	2028	0.786	0.905
8	2029	0.759	0.893
9	2030	0.734	0.881
10	2031	0.709	0.870
11	2032	0.685	0.858
12	2033	0.662	0.847
13	2034	0.639	0.837
14	2035	0.618	0.826
15	2036	0.597	0.816
16	2037	0.577	0.806
17	2038	0.557	0.797
18	2039	0.538	0.787
19	2040	0.520	0.778
20	2041	0.503	0.769
21	2042	0.486	0.760
22	2043	0.469	0.752
23	2044	0.453	0.743
24	2045	0.438	0.735
25	2046	0.423	0.727
26	2047	0.409	0.719
27	2048	0.395	0.712
28	2049	0.382	0.704
29	2050	0.369	0.697
30	2051	0.356	0.690

11.5.1 Evaluation of Alternative Performance Across Criteria

The following tables provide a comprehensive summary of the weighted performance of each alternative across different criteria, applying an intertemporal weighting coefficient on a yearly basis to ensure consistency in the evaluation process. These

Table 11.23 Performance values of alternatives for the "budget balance" criterion

Budget balance							
t	Period	Year	Alternative 0	GoFooodLog	Safe and secure parking	ZLS	Intermodal terminal
0	Present	2021	1.458	1.458	1.458	1.458	1.458
1	Short-term	2022	1.409	1.409	1.409	1.409	1.409
2	Short-term	2023	1.361	1.361	1.361	1.361	1.361
3	Short-term	2024	1.315	1.315	1.315	1.315	1.315
4	Medium-term	2025	1.271	1.271	1.271	1.271	1.271
5	Medium-term	2026	1.228	1.228	1.228	1.228	1.228
6	Medium-term	2027	1.186	1.186	1.186	1.186	1.186
7	Medium-term	2028	1.146	1.146	1.146	1.146	1.146
8	Medium-term	2029	1.107	1.107	1.107	1.107	1.107
9	Medium-term	2030	1.070	1.070	1.070	1.070	1.070
10	Medium-term	2031	1.034	1.034	1.034	1.034	1.034
11	Long-term	2032	0.999	0.999	0.999	0.999	0.999
12	Long-term	2033	0.965	0.965	0.965	0.965	0.965
13	Long-term	2034	0.932	0.932	0.932	0.932	0.932
14	Long-term	2035	0.901	0.901	0.901	0.901	0.901
15	Long-term	2036	0.870	0.870	0.870	0.870	0.870
16	Long-term	2037	0.841	0.841	0.841	0.841	0.841
17	Long-term	2038	0.813	0.813	0.813	0.813	0.813
18	Long-term	2039	0.785	0.785	0.785	0.785	0.785
19	Long-term	2040	0.759	0.759	0.759	0.759	0.759
20	Long-term	2041	0.733	0.733	0.733	0.733	0.733
21	Long-term effects	2042	0.708	0.708	0.708	0.708	0.708
22	Long-term effects	2043	0.684	0.684	0.684	0.684	0.684
23	Long-term effects	2044	0.661	0.661	0.661	0.661	0.661
24	Long-term effects	2045	0.639	0.639	0.639	0.639	0.639

(continued)

Table 11.23 (continued)

Budget balance							
t	Period	Year	Alternative 0	GoFooodLog	Safe and secure parking	ZLS	Intermodal terminal
25	Long-term effects	2046	0.617	0.617	0.617	0.617	0.617
26	Long-term effects	2047	0.596	0.596	0.596	0.596	0.596
27	Long-term effects	2048	0.576	0.576	0.576	0.576	0.576
28	Long-term effects	2049	0.557	0.557	0.557	0.557	0.557
29	Long-term effects	2050	0.538	0.538	0.538	0.538	0.538
30	Long-term effects	2051	0.520	0.520	0.520	0.520	0.520

tables integrate the short-, medium-, and long-term performance evaluations for SDAG (Table 11.23—Short-Term Performance for SDAG, Table 11.24—Medium-Term Performance for SDAG, Table 11.25—Long-Term Performance for SDAG), with the appropriate temporal weighting coefficients as outlined in Table 11.22—Temporal Horizon and Intertemporal Weighting Coefficients.

Table 11.23 summarizes the weighted performance results for each alternative under the budget balance criterion, with the yearly application of the τ_r^e coefficient, given its relevance to monetary aspects.

Table 11.24 presents the weighted performance results for each alternative under the economic incentives criterion, applying the τ_r^e coefficient annually, as this criterion is also linked to financial considerations.

Table 11.25 outlines the weighted performance results for each alternative concerning the overcoming marginalization criterion, with the $\tau_r{}^e$ coefficient applied on an annual basis due to its monetary implications.

Table 11.26 details the weighted performance outcomes for each alternative regarding the expansion of SDAG's activities, again incorporating the $\tau_r{}^e$ coefficient annually, given its financial significance.

Table 11.27 reports the weighted performance results for each alternative under the establishment of new activities within SDAG criterion, with the τ_r^e coefficient applied year by year due to its monetary relevance.

Unlike the previous tables, Table 11.28 evaluates alternatives based on their contribution to economic development in the region, a criterion linked to broader societal well-being rather than direct monetary value. Therefore, the $\tau_r{}^h$ coefficient is applied, reflecting its relevance to social and non-monetizable benefits.

By integrating the data from Tables 11.23, 11.24, 11.25, 11.26, 11.27 to 11.28, it is possible to construct performance matrices for each alternative, adjusted and weighted for each temporal phase. These matrices provide a holistic overview of the alternatives' long-term viability.

Furthermore, for each key stakeholder, a dedicated table will present the updated annual performance of alternatives. These tables are derived by combining the individual annual performance values of each alternative across all criteria, which were previously detailed only for SDAG. This stakeholder-specific analysis enhances the contextual relevance of the results and supports more informed decision-making processes.

11.5.1.1 SDAG

By aggregating the values first across the various intervals and then into a single result, the following Table 11.29 is obtained.

By first aggregating the values within the different intervals and then consolidating them into a single result, the following table is obtained (Tables 11.30, 11.31 and Figs. 11.21, 11.22).

It is observed that for SDAG, the zero alternative is the least effective, while the GoFoodLog alternative yields the most satisfactory results. Considering the weight of the temporal value, an increase in the distance between the values becomes evident,

Table 11.24 Performance values of alternatives for the "economic incentives" criterion

Economic incentives							
t	Period	Year	Alternative 0	GoFooodLog	Safe and secure parking	ZLS	Intermodal terminal
0	Present	2021	0.000	0.000	0.000	0.000	0.000
1	Short-term	2022	0.000	0.000	0.000	0.000	0.000
2	Short-term	2023	0.000	0.000	0.000	0.000	0.000
3	Short-term	2024	0.000	0.564	0.564	0.564	0.564
4	Medium-term	2025	0.000	0.545	0.545	0.545	0.545
5	Medium-term	2026	0.000	0.526	0.526	0.526	0.526
6	Medium-term	2027	0.000	0.508	0.508	0.508	0.508
7	Medium-term	2028	0.000	0.491	0.491	0.491	0.491
8	Medium-term	2029	0.000	0.475	0.475	0.475	0.475
9	Medium-term	2030	0.000	0.459	0.459	0.459	0.459
10	Medium-term	2031	0.000	0.443	0.443	0.443	0.443
11	Long-term	2032	0.000	0.428	0.428	0.428	0.428
12	Long-term	2033	0.000	0.414	0.414	0.414	0.414
13	Long-term	2034	0.000	0.400	0.400	0.400	0.400
14	Long-term	2035	0.000	0.386	0.386	0.386	0.386

(continued)

Table 11.24 (continued)

Economic incentives							
t	Period	Year	Alternative 0	GoFooodLog	Safe and secure parking	ZLS	Intermodal terminal
15	Long-term	2036	0.000	0.373	0.373	0.373	0.373
16	Long-term	2037	0.000	0.360	0.360	0.360	0.360
17	Long-term	2038	0.000	0.348	0.348	0.348	0.348
18	Long-term	2039	0.000	0.336	0.336	0.336	0.336
19	Long-term	2040	0.000	0.325	0.325	0.325	0.325
20	Long-term	2041	0.000	0.314	0.314	0.314	0.314
21	Long-term effects	2042	0.000	0.303	0.303	0.303	0.303
22	Long-term effects	2043	0.000	0.293	0.293	0.293	0.293
23	Long-term effects	2044	0.000	0.283	0.283	0.283	0.283
24	Long-term effects	2045	0.000	0.274	0.274	0.274	0.274
25	Long-term effects	2046	0.000	0.264	0.264	0.264	0.264
26	Long-term effects	2047	0.000	0.256	0.256	0.256	0.256
27	Long-term effects	2048	0.000	0.247	0.247	0.247	0.247
28	Long-term effects	2049	0.000	0.239	0.239	0.239	0.239
29	Long-term effects	2050	0.000	0.230	0.230	0.230	0.230
30	Long-term effects	2051	0.000	0.223	0.223	0.223	0.223

Table 11.25 Performance values of alternatives for the "overcoming marginality" criterion

Overcoming marginality							
t	Period	Year	Alternative 0	GoFooodLog	Safe and secure parking	ZLS	Intermodal terminal
0	Present	2021	0.000	0.000	0.000	0.000	0.000
1	Short-term	2022	0.000	0.000	0.000	0.000	0.000
2	Short-term	2023	0.000	0.000	0.000	0.000	0.000

(continued)

Table 11.25 (continued)

Overcoming marginality							
t	Period	Year	Alternative 0	GoFooodLog	Safe and secure parking	ZLS	Intermodal terminal
3	Short-term	2024	0.000	0.902	1.127	0.676	0.000
4	Medium-term	2025	0.000	0.871	1.089	0.654	0.000
5	Medium-term	2026	0.000	0.842	1.052	0.631	0.000
6	Medium-term	2027	0.000	0.814	1.017	0.610	0.000
7	Medium-term	2028	0.000	0.786	0.982	0.589	0.000
8	Medium-term	2029	0.000	0.759	0.949	0.570	0.000
9	Medium-term	2030	0.000	0.734	0.917	0.550	0.000
10	Medium-term	2031	0.000	0.709	0.532	0.532	0.354
11	Long-term	2032	0.000	0.685	0.514	0.514	0.342
12	Long-term	2033	0.000	0.662	0.496	0.496	0.331
13	Long-term	2034	0.000	0.639	0.480	0.480	0.320
14	Long-term	2035	0.000	0.618	0.463	0.463	0.309
15	Long-term	2036	0.000	0.597	0.448	0.448	0.298
16	Long-term	2037	0.000	0.577	0.433	0.433	0.288
17	Long-term	2038	0.000	0.557	0.418	0.418	0.279
18	Long-term	2039	0.000	0.538	0.404	0.404	0.269
19	Long-term	2040	0.000	0.520	0.390	0.390	0.260
20	Long-term	2041	0.000	0.377	0.377	0.377	0.377
21	Long-term effects	2042	0.000	0.364	0.364	0.364	0.364
22	Long-term effects	2043	0.000	0.352	0.352	0.352	0.352
23	Long-term effects	2044	0.000	0.340	0.340	0.340	0.340
24	Long-term effects	2045	0.000	0.328	0.328	0.328	0.328
25	Long-term effects	2046	0.000	0.317	0.317	0.317	0.317
26	Long-term effects	2047	0.000	0.307	0.307	0.307	0.307
27	Long-term effects	2048	0.000	0.296	0.296	0.296	0.296
28	Long-term effects	2049	0.000	0.286	0.286	0.286	0.286
29	Long-term effects	2050	0.000	0.277	0.277	0.277	0.277
30	Long-term effects	2051	0.000	0.267	0.267	0.267	0.267

highlighting that the alternative which maximizes the SDAG criteria the most is GoFoodLog. The performance of the different time periods can also be compared, although no significant upheavals emerge. For SDAG, the preferable alternative remains GoFoodLog.

Table 11.26 Performance values of alternatives for the "increase in activities performed by SDAG" criterion

Increase in activities Performed by SDAG							
t	Period	Year	Alternative 0	GoFooodLog	Safe and secure parking	ZLS	Intermodal terminal
0	Present	2021	0.000	0.000	0.000	0.000	0.000
1	Short-term	2022	0.000	0.000	0.000	0.000	0.000
2	Short-term	2023	0.000	0.000	0.000	0.000	0.000
3	Short-term	2024	0.000	0.225	0.075	0.000	0.000
4	Medium-term	2025	0.000	0.218	0.073	0.000	0.000
5	Medium-term	2026	0.000	0.210	0.070	0.000	0.000
6	Medium-term	2027	0.000	0.203	0.068	0.000	0.000
7	Medium-term	2028	0.000	0.196	0.065	0.000	0.000
8	Medium-term	2029	0.000	0.190	0.063	0.000	0.000
9	Medium-term	2030	0.000	0.183	0.061	0.000	0.000
10	Medium-term	2031	0.000	0.236	0.059	0.000	0.059
11	Long-term	2032	0.000	0.228	0.057	0.000	0.057
12	Long-term	2033	0.000	0.221	0.055	0.000	0.055
13	Long-term	2034	0.000	0.213	0.053	0.000	0.053
14	Long-term	2035	0.000	0.206	0.051	0.000	0.051
15	Long-term	2036	0.000	0.199	0.050	0.000	0.050
16	Long-term	2037	0.000	0.192	0.048	0.000	0.048
17	Long-term	2038	0.000	0.186	0.046	0.000	0.046
18	Long-term	2039	0.000	0.179	0.045	0.000	0.045
19	Long-term	2040	0.000	0.173	0.043	0.000	0.043
20	Long-term	2041	0.000	0.126	0.042	0.000	0.084
21	Long-term effects	2042	0.000	0.121	0.040	0.000	0.081

(continued)

Table 11.26 (continued)

Increase in activities Performed by SDAG							
t	Period	Year	Alternative 0	GoFooodLog	Safe and secure parking	ZLS	Intermodal terminal
22	Long-term effects	2043	0.000	0.117	0.039	0.000	0.078
23	Long-term effects	2044	0.000	0.113	0.038	0.000	0.076
24	Long-term effects	2045	0.000	0.109	0.036	0.000	0.073
25	Long-term effects	2046	0.000	0.106	0.035	0.000	0.071
26	Long-term effects	2047	0.000	0.102	0.034	0.000	0.068
27	Long-term effects	2048	0.000	0.099	0.033	0.000	0.066
28	Long-term effects	2049	0.000	0.095	0.032	0.000	0.064
29	Long-term effects	2050	0.000	0.092	0.031	0.000	0.061
30	Long-term effects	2051	0.000	0.089	0.030	0.000	0.059

Table 11.27 Performance values of alternatives for the "establishment of new clients in SDAG" criterion

Establishment of new clients in SDAG							
t	Period	Year	Alternative 0	GoFooodLog	Safe and secure parking	ZLS	Intermodal terminal
0	Present	2021	0.000	0.000	0.000	0.000	0.000
1	Short-term	2022	0.000	0.000	0.000	0.000	0.000
2	Short-term	2023	0.000	0.000	0.000	0.000	0.000
3	Short-term	2024	0.000	0.752	0.188	0.376	0.000
4	Medium-term	2025	0.000	0.726	0.182	0.363	0.000
5	Medium-term	2026	0.000	0.702	0.175	0.351	0.000
6	Medium-term	2027	0.000	0.678	0.169	0.339	0.000

(continued)

Table 11.27 (continued)

Establishment of new clients in SDAG							
t	Period	Year	Alternative 0	GoFooodLog	Safe and secure parking	ZLS	Intermodal terminal
7	Medium-term	2028	0.000	0.655	0.164	0.327	0.000
8	Medium-term	2029	0.000	0.633	0.158	0.316	0.000
9	Medium-term	2030	0.000	0.611	0.153	0.306	0.000
10	Medium-term	2031	0.000	0.738	0.443	0.591	0.148
11	Long-term	2032	0.000	0.713	0.428	0.571	0.143
12	Long-term	2033	0.000	0.689	0.414	0.551	0.138
13	Long-term	2034	0.000	0.666	0.400	0.533	0.133
14	Long-term	2035	0.000	0.644	0.386	0.515	0.129
15	Long-term	2036	0.000	0.622	0.373	0.497	0.124
16	Long-term	2037	0.000	0.601	0.360	0.481	0.120
17	Long-term	2038	0.000	0.580	0.348	0.464	0.116
18	Long-term	2039	0.000	0.561	0.336	0.449	0.112
19	Long-term	2040	0.000	0.542	0.325	0.433	0.108
20	Long-term	2041	0.000	0.209	0.314	0.314	0.209
21	Long-term effects	2042	0.000	0.202	0.303	0.303	0.202
22	Long-term effects	2043	0.000	0.195	0.293	0.293	0.195
23	Long-term effects	2044	0.000	0.189	0.283	0.283	0.189
24	Long-term effects	2045	0.000	0.182	0.274	0.274	0.182
25	Long-term effects	2046	0.000	0.176	0.264	0.264	0.176
26	Long-term effects	2047	0.000	0.170	0.256	0.256	0.170
27	Long-term effects	2048	0.000	0.165	0.247	0.247	0.165
28	Long-term effects	2049	0.000	0.159	0.239	0.239	0.159
29	Long-term effects	2050	0.000	0.154	0.230	0.230	0.154
30	Long-term effects	2051	0.000	0.148	0.223	0.223	0.148

11.5.1.2 Municipality of Gorizia

By first aggregating the values within the different intervals and then consolidating them into a single result, the following table is obtained (Tables 11.32, 11.33, 11.34 and Figs. 11.23, 11.24).

For the Municipality of Gorizia, the results are similar to those of SDAG, albeit with lower values. The zero alternative is entirely ineffective, and the most significant difference is found in the Safe and Secure Parking alternative, which does not benefit the territory but only the company managing the spaces. For SDAG, this alternative ranks second, whereas for the Municipality of Gorizia, it ranks last.

Table 11.28 Performance values of alternatives for the "economic development of the territory" criterion

Economic development of the territory							
t	Period	Year	Alternative 0	GoFooodLog	Safe and secure parking	ZLS	Intermodal terminal
0	Present	2021	0.000	0.000	0.000	0.000	0.000
1	Short-term	2022	0.000	0.000	0.000	0.000	0.000
2	Short-term	2023	0.000	0.000	0.000	0.000	0.000
3	Short-term	2024	0.000	0.075	0.038	0.075	0.000
4	Medium-term	2025	0.000	0.073	0.036	0.073	0.000
5	Medium-term	2026	0.000	0.070	0.035	0.070	0.000
6	Medium-term	2027	0.000	0.068	0.034	0.068	0.000
7	Medium-term	2028	0.000	0.065	0.033	0.065	0.000
8	Medium-term	2029	0.000	0.063	0.032	0.063	0.000
9	Medium-term	2030	0.000	0.061	0.031	0.061	0.000
10	Medium-term	2031	0.000	0.089	0.030	0.089	0.030
11	Long-term	2032	0.000	0.086	0.029	0.086	0.029
12	Long-term	2033	0.000	0.083	0.028	0.083	0.028
13	Long-term	2034	0.000	0.080	0.027	0.080	0.027
14	Long-term	2035	0.000	0.077	0.026	0.077	0.026
15	Long-term	2036	0.000	0.075	0.025	0.075	0.025
16	Long-term	2037	0.000	0.072	0.024	0.072	0.024
17	Long-term	2038	0.000	0.070	0.023	0.070	0.023
18	Long-term	2039	0.000	0.067	0.022	0.067	0.022
19	Long-term	2040	0.000	0.065	0.022	0.065	0.022
20	Long-term	2041	0.000	0.074	0.031	0.096	0.042
21	Long-term effects	2042	0.000	0.072	0.030	0.095	0.000

(continued)

Table 11.28 (continued)

Economic development of the territory							
t	Period	Year	Alternative 0	GoFooodLog	Safe and secure parking	ZLS	Intermodal terminal
22	Long-term effects	2043	0.000	0.071	0.030	0.094	0.000
23	Long-term effects	2044	0.000	0.069	0.030	0.093	0.000
24	Long-term effects	2045	0.000	0.068	0.029	0.092	0.000
25	Long-term effects	2046	0.000	0.067	0.029	0.091	0.000
26	Long-term effects	2047	0.000	0.065	0.028	0.090	0.000
27	Long-term effects	2048	0.000	0.064	0.028	0.089	0.000
28	Long-term effects	2049	0.000	0.063	0.027	0.088	0.000
29	Long-term effects	2050	0.000	0.062	0.027	0.087	0.000
30	Long-term effects	2051	0.000	0.061	0.027	0.086	0.030

It is noted that through the long-term actualization, there is a change in the classification of alternatives for the Municipality of Gorizia: the alternative that maximizes the criteria the most becomes GoFoodLog, whereas for the non-actualized values, it was the Simplified Logistics Zone.

11.5.1.3 Port System Authority

By first aggregating the values within the different intervals and then consolidating them into a single result, the following table is obtained (Tables 11.35, 11.36, 11.37 and Figs. 11.25, 11.26).

Considering the weight of the temporal value, the results differ only slightly from the non-actualized values, with GoFoodLog being the alternative that maximizes the criteria. However, it is noted that with actualization, the Intermodal Terminal alternative is the most penalized: it remains in last place for all time periods except 2021–2051, where it surpasses the zero alternative. The Intermodal Terminal does not even surpass the Safe and Secure Park alternative, which only occurred starting from the long-term effects, whose incidence is significantly reduced by actualization. This is because the Intermodal Terminal would only begin to perform well once its construction is fully completed, which will only happen far into the future.

Table 11.29 Year-by-year updated performance of alternatives for SDAG

SDAG							
t	Period	Year	Alternative 0	GoFooodLog	Safe and secure parking	ZLS	Intermodal terminal
0	Present	2021	1.46	1.46	1.46	1.46	1.46
1	Short-term	2022	1.41	1.41	1.41	1.41	1.41
2	Short-term	2023	1.36	1.36	1.36	1.36	1.36
3	Short-term	2024	1.32	3.83	3.31	3.01	1.88
4	Medium-term	2025	1.27	3.70	3.20	2.90	1.82
5	Medium-term	2026	1.23	3.58	3.09	2.81	1.75
6	Medium-term	2027	1.19	3.46	2.98	2.71	1.69
7	Medium-term	2028	1.15	3.34	2.88	2.62	1.64
8	Medium-term	2029	1.11	3.23	2.78	2.53	1.58
9	Medium-term	2030	1.07	3.12	2.69	2.45	1.53
10	Medium-term	2031	1.03	3.25	2.54	2.69	2.07
11	Long-term	2032	1.00	3.14	2.45	2.60	2.00
12	Long-term	2033	0.97	3.03	2.37	2.51	1.93
13	Long-term	2034	0.93	2.93	2.29	2.42	1.86
14	Long-term	2035	0.90	2.83	2.21	2.34	1.80
15	Long-term	2036	0.87	2.74	2.14	2.26	1.74
16	Long-term	2037	0.84	2.64	2.07	2.19	1.68
17	Long-term	2038	0.81	2.55	2.00	2.11	1.63
18	Long-term	2039	0.79	2.47	1.93	2.04	1.57
19	Long-term	2040	0.76	2.38	1.86	1.97	1.52
20	Long-term	2041	0.73	1.83	1.81	1.83	1.76
21	Long-term effects	2042	0.71	1.77	1.75	1.77	1.66
22	Long-term effects	2043	0.68	1.71	1.69	1.72	1.60
23	Long-term effects	2044	0.66	1.66	1.63	1.66	1.55
24	Long-term effects	2045	0.64	1.60	1.58	1.61	1.50
25	Long-term effects	2046	0.62	1.55	1.53	1.55	1.45
26	Long-term effects	2047	0.60	1.50	1.48	1.50	1.40
27	Long-term effects	2048	0.58	1.45	1.43	1.46	1.35
28	Long-term effects	2049	0.56	1.40	1.38	1.41	1.30

(continued)

Table 11.29 (continued)

SDAG							
t	Period	Year	Alternative 0	GoFooodLog	Safe and secure parking	ZLS	Intermodal terminal
29	Long-term effects	2050	0.54	1.35	1.33	1.36	1.26
30	Long-term effects	2051	0.52	1.31	1.29	1.32	1.25

Table 11.30 Performance of the alternatives across different timeframe intervals for SDAG

Updated performance for SDAG					
Timeframe intervals	Alternatives				
	Alternative 0	GoFooodLog	Safe and secure parking	ZLS	Intermodal terminal
2021–2024	5.544	8.062	7.536	7.235	6.108
2025–2031	8.043	28.701	26.472	22.355	12.405
2032–2041	8.598	26.551	21.137	22.284	17.489
2042–2051	6.095	15.291	15.088	15.360	14.310
2021–2051	28.3	78.6	70.2	67.2	50.3

Table 11.31 Performance of the alternatives across different timeframe periods for SDAG

Updated performance for SDAG					
Timeframe periods	Alternatives				
	Alternative 0	GoFooodLog	Safe and secure parking	ZLS	Intermodal terminal
2021–2024	5.544	8.062	7.536	7.235	6.108
2021–2031	13.587	31.737	27.698	25.943	18.188
2021–2041	22.185	58.288	48.835	48.227	35.677
2021–2051	28.280	73.580	63.923	63.587	49.987

11.5.1.4 Friuli Venezia Giulia Region

By first aggregating the values within the different intervals and then consolidating them into a single result, the following table is obtained (Tables 11.38, 11.39, 11.40 and Figs. 11.27, 11.28).

Considering the weight of the temporal value for the Friuli Venezia Giulia Region, the most compelling alternative remains the Simplified Logistics Zone, which demonstrates the best and most consistent performance over time. In the first two time periods, the Intermodal Terminal represents the least attractive alternative. However,

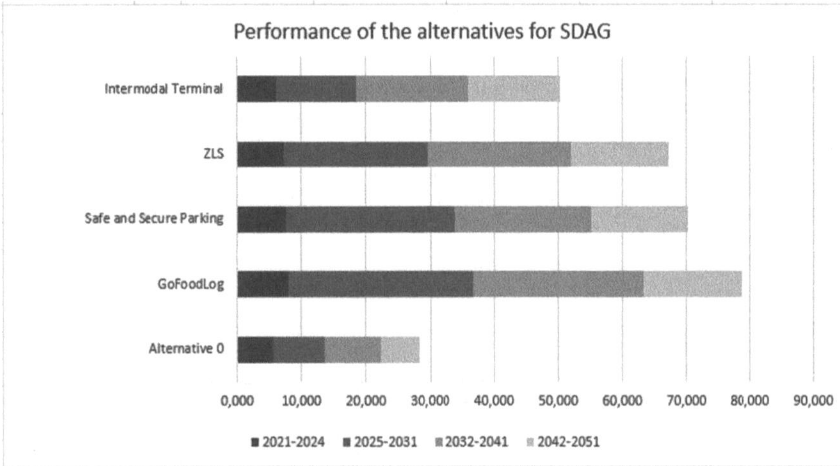

Fig. 11.21 Histogram of the performance of the alternatives in the different time intervals for SDAG, referring to Table 11.30

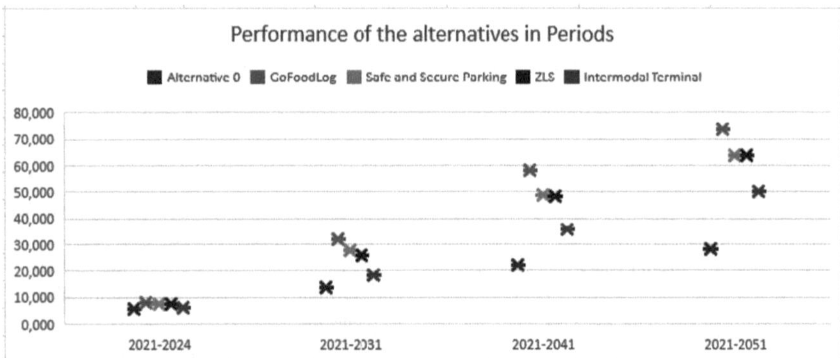

Fig. 11.22 Graph of comparison of the performance of the alternatives in the different time spans for SDAG, referring to Table 11.31

from the third period onwards, its performance improves sufficiently to surpass the zero alternative in the fourth period.

11.5.2 Results

The matrix of alternative values for the stakeholders for the interval 2021–2024 is as follows (Table 11.41).

Table 11.32 Year-by-year updated performance of alternatives for Municipality of Gorizia

Municipality of Gorizia							
t	Period	Year	Alternative 0	GoFooodLog	Safe and secure parking	ZLS	Intermodal terminal
0	Present	2021	0.0	0.0	0.0	0.0	0.0
1	Short-term	2022	0.0	0.0	0.0	0.0	0.0
2	Short-term	2023	0.0	0.0	0.0	0.0	0.0
3	Short-term	2024	0.0	2.9	0.5	2.3	0.0
4	Medium-term	2025	0.0	2.8	0.4	2.3	0.0
5	Medium-term	2026	0.0	2.8	0.4	2.2	0.0
6	Medium-term	2027	0.0	2.7	0.4	2.1	0.0
7	Medium-term	2028	0.0	2.6	0.4	2.1	0.0
8	Medium-term	2029	0.0	2.6	0.4	2.0	0.0
9	Medium-term	2030	0.0	2.5	0.4	2.0	0.0
10	Medium-term	2031	0.0	2.5	0.4	2.6	0.8
11	Long-term	2032	0.0	2.4	0.3	2.6	0.8
12	Long-term	2033	0.0	2.4	0.3	2.5	0.8
13	Long-term	2034	0.0	2.3	0.3	2.4	0.7
14	Long-term	2035	0.0	2.3	0.3	2.4	0.7
15	Long-term	2036	0.0	2.2	0.3	2.3	0.7
16	Long-term	2037	0.0	2.2	0.3	2.3	0.7
17	Long-term	2038	0.0	2.1	0.3	2.2	0.7
18	Long-term	2039	0.0	2.1	0.3	2.2	0.7
19	Long-term	2040	0.0	2.0	0.3	2.1	0.6
20	Long-term	2041	0.0	2.0	0.3	2.1	1.9
21	Long-term effects	2042	0.0	2.0	0.2	2.0	1.9
22	Long-term effects	2043	0.0	1.9	0.2	2.0	1.8
23	Long-term effects	2044	0.0	1.9	0.2	1.9	1.8
24	Long-term effects	2045	0.0	1.9	0.2	1.9	1.8
25	Long-term effects	2046	0.0	1.8	0.2	1.8	1.7
26	Long-term effects	2047	0.0	1.8	0.2	1.8	1.7
27	Long-term effects	2048	0.0	1.7	0.2	1.8	1.7
28	Long-term effects	2049	0.0	1.7	0.2	1.7	1.6

(continued)

Table 11.32 (continued)

Municipality of Gorizia							
t	Period	Year	Alternative 0	GoFooodLog	Safe and secure parking	ZLS	Intermodal terminal
29	Long-term effects	2050	0.0	1.7	0.2	1.7	1.6
30	Long-term effects	2051	0.0	1.7	0.2	1.7	1.6

Table 11.33 Performance of the alternatives across different time intervals for Municipality of Gorizia

Updated performance for Municipality of Gorizia					
Timeframe	Alternatives				
	Alternative 0	GoFooodLog	Safe and secure parking	ZLS	Intermodal terminal
2021–2024	0.000	2.908	0.451	2.310	0.000
2025–2031	0.000	18.574	2.757	15.293	0.789
2032–2041	0.000	22.044	2.948	22.961	8.281
2042–2051	0.000	18.035	2.090	18.315	17.130
2021–2051	0.0	61.6	8.2	58.9	26.2

Table 11.34 Performance of the alternatives across different time period for Municipality of Gorizia

Updated performance for SDAG					
Timeframe periods	Alternatives				
	Alternative 0	GoFooodLog	Safe and secure parking	ZLS	Intermodal terminal
2021–2024	0.000	2.908	0.451	2.310	0.000
2021–2031	0.000	21.482	3.208	17.602	0.789
2021–2041	0.000	43.527	6.156	40.564	9.071
2021–2051	0.000	61.562	8.246	58.879	26.201

Actor/Alternative matrix for the time interval 2021–2024, with weighted and actualized values, does not show substantial differences from the short-term matrix, except for the values themselves. The preferred alternative for most stakeholders remains GoFoodLog, except for the Friuli Venezia Giulia Region, which prefers the Simplified Logistics Zone (Fig. 11.29).

The matrix of alternative values for the stakeholders for the interval 2025–2031 is as follows (Table 11.42).

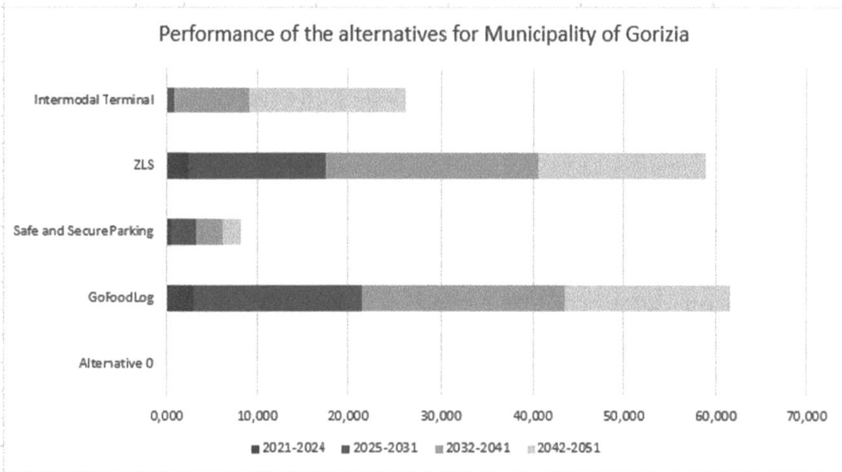

Fig. 11.23 Histogram of the performance of the alternatives in the different time intervals for Municipality of Gorizia, referring to Table 11.33

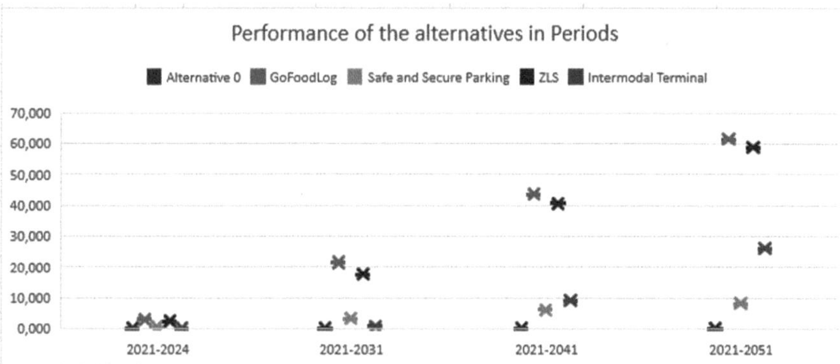

Fig. 11.24 Graph of comparison of the performance of the alternatives in the different time spans for Municipality of Gorizia, referring to Table 11.34

The Actor/Alternative matrix for the time interval 2025–2031, with weighted and actualized values, does not change the balance of preferences compared to the first interval: the preferred alternative remains GoFoodLog for most stakeholders (Fig. 11.30).

Incorporating the values from the previous matrix to evaluate the results over the time span 2021–2031, the preferences remain the same, with only slight changes in values. The matrix of alternative values for the stakeholders for the time span 2021–2031 is as follows. The matrix of alternative values for the stakeholders for the time interval 2032–2041 is as follows (Table 11.43 and Fig. 11.31).

Table 11.35 Year-by-year updated performance of alternatives for Port System Authority

Port System Authority							
t	Period	Year	Alternative 0	GoFooodLog	Safe and secure parking	ZLS	Intermodal terminal
0	Present	2021	1.1	1.1	1.1	1.1	1.1
1	Short-term	2022	1.1	1.1	1.1	1.1	1.1
2	Short-term	2023	1.0	1.0	1.0	1.0	1.0
3	Short-term	2024	1.0	3.1	1.2	2.6	0.0
4	Medium-term	2025	1.0	3.0	1.2	2.6	0.0
5	Medium-term	2026	0.9	2.9	1.2	2.5	0.0
6	Medium-term	2027	0.9	2.9	1.1	2.5	0.0
7	Medium-term	2028	0.9	2.8	1.1	2.4	0.0
8	Medium-term	2029	0.9	2.7	1.1	2.4	0.0
9	Medium-term	2030	0.8	2.6	1.0	2.3	0.0
10	Medium-term	2031	0.2	2.5	1.0	2.1	0.7
11	Long-term	2032	0.2	2.4	1.0	2.0	0.7
12	Long-term	2033	0.2	2.3	1.0	2.0	0.7
13	Long-term	2034	0.2	2.2	0.9	1.9	0.7
14	Long-term	2035	0.2	2.2	0.9	1.8	0.6
15	Long-term	2036	0.2	2.1	0.9	1.8	0.6
16	Long-term	2037	0.2	2.0	0.8	1.7	0.6
17	Long-term	2038	0.2	2.0	0.8	1.7	0.6
18	Long-term	2039	0.2	1.9	0.8	1.6	0.6
19	Long-term	2040	0.2	1.8	0.8	1.6	0.5
20	Long-term	2041	0.3	1.8	0.8	1.9	1.1
21	Long-term effects	2042	0.3	1.7	0.7	1.9	0.8
22	Long-term effects	2043	0.3	1.7	0.7	1.9	0.8
23	Long-term effects	2044	0.3	1.7	0.7	1.8	0.8
24	Long-term effects	2045	0.3	1.6	0.7	1.8	0.8
25	Long-term effects	2046	0.3	1.6	0.7	1.8	0.7
26	Long-term effects	2047	0.3	1.6	0.6	1.8	0.7
27	Long-term effects	2048	0.3	1.5	0.6	1.7	0.7
28	Long-term effects	2049	0.3	1.5	0.6	1.7	0.7

(continued)

Table 11.35 (continued)

Port System Authority							
t	Period	Year	Alternative 0	GoFooodLog	Safe and secure parking	ZLS	Intermodal terminal
29	Long-term effects	2050	0.3	1.5	0.6	1.7	0.7
30	Long-term effects	2051	0.3	1.4	0.6	1.6	0.8

Table 11.36 Performance of the alternatives across different time intervals for Port System Authority

Updated performance for Port System Authority					
Timeframe	Alternatives				
	Alternative 0	GoFooodLog	Safe and secure parking	ZLS	Intermodal terminal
2021–2024	4.167	6.306	4.418	5.810	3.174
2025–2031	5.601	19.406	7.745	16.783	0.724
2032–2041	2.285	20.777	8.647	17.993	6.717
2042–2051	2.633	15.918	6.534	17.724	7.590
2021–2051	14.7	62.4	27.3	58.3	18.2

Table 11.37 Performance of the alternatives across different time periods for Port System Authority

Updated performance for Port System Authority					
Timeframe periods	Alternatives				
	Alternative 0	GoFooodLog	Safe and secure parking	ZLS	Intermodal terminal
2021–2024	4.167	6.306	4.418	5.810	3.174
2021–2031	9.768	25.712	12.164	22.593	3.897
2021–2041	12.054	46.488	20.811	40.586	10.614
2021–2051	14.687	62.406	27.345	58.310	18.204

The matrix of alternative values for the stakeholders for the interval 2032–2041 is as follows (Table 11.44).

Regarding the Actor/Alternative matrix for the interval 2032–2041, there is a realignment with the long-term matrix (non-actualized values), showing an equal preference for GoFoodLog (SDAG and Port System Authority) and the Enhanced Simplified Logistics Zone (Municipality of Gorizia and Friuli Venezia Giulia Region) (Fig. 11.32).

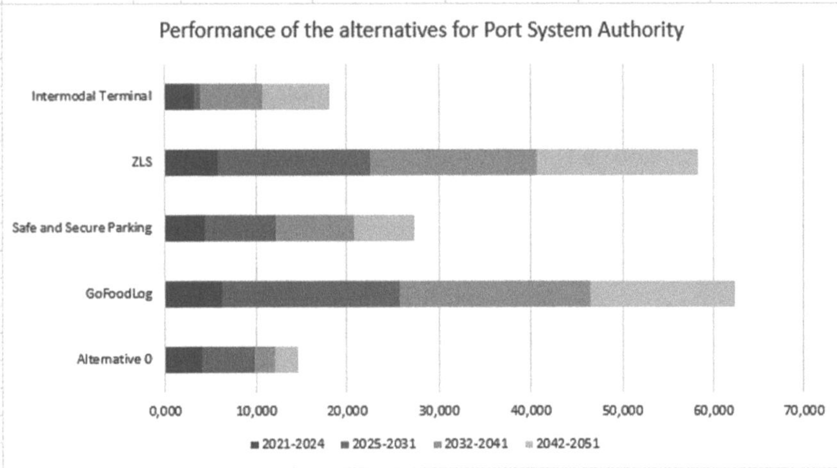

Fig. 11.25 Histogram of the performance of the alternatives in the different time intervals for Port System Authority, referring to Table 11.36

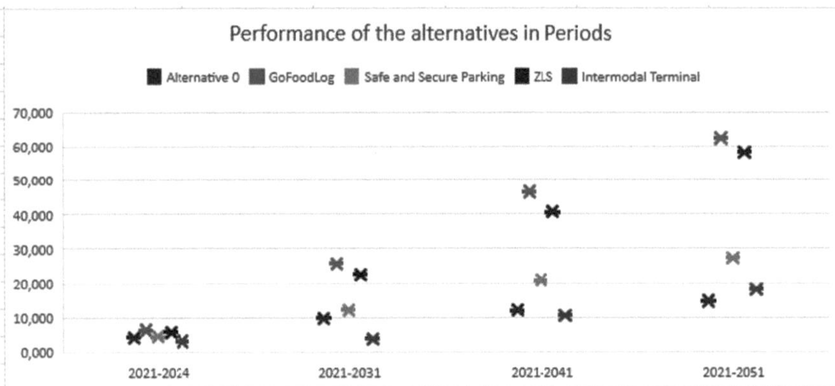

Fig. 11.26 Graph of comparison of the performance of the alternatives in the different time periods for Port System Authority, referring to Table 11.37

Considering the performance over the entire time period 2021–2041 (incorporating previous results), there is again a change in the preference of the Municipality of Gorizia, and thus a shift in the balance, with the Friuli Venezia Giulia Region remaining the only one preferring the Simplified Logistics Zone (Table 11.45 and Fig. 11.33).

The matrix of alternative values for the stakeholders for the interval 2042–2051 is as follows (Table 11.46).

The Actor/Alternative matrix for the time interval 2042–2051, with weighted and actualized values, highlights a further shift in balance, with GoFoodLog becoming

Table 11.38 Year-by-year updated performance of alternatives for Friuli Venezia Giulia Region

SDAG

t	Period	Year	Alternative 0	GoFooodLog	Safe and secure parking	ZLS	Intermodal terminal
0	Present	2021	0.7	0.7	0.7	0.7	0.7
1	Short-term	2022	0.7	0.7	0.7	0.7	0.7
2	Short-term	2023	0.7	0.7	0.7	0.7	0.7
3	Short-term	2024	0.7	2.2	1.2	3.2	0.0
4	Medium-term	2025	0.6	2.1	1.2	3.1	0.0
5	Medium-term	2026	0.6	2.1	1.1	3.0	0.0
6	Medium-term	2027	0.6	2.0	1.1	2.9	0.0
7	Medium-term	2028	0.6	1.9	1.1	2.8	0.0
8	Medium-term	2029	0.6	1.9	1.0	2.7	0.0
9	Medium-term	2030	0.6	1.8	1.0	2.6	0.0
10	Medium-term	2031	0.2	2.1	1.3	2.5	0.7
11	Long-term	2032	0.2	2.0	1.2	2.5	0.7
12	Long-term	2033	0.2	2.0	1.2	2.4	0.7
13	Long-term	2034	0.2	1.9	1.1	2.3	0.7
14	Long-term	2035	0.2	1.9	1.1	2.2	0.6
15	Long-term	2036	0.2	1.8	1.1	2.1	0.6
16	Long-term	2037	0.2	1.7	1.0	2.1	0.6
17	Long-term	2038	0.2	1.7	1.0	2.0	0.6
18	Long-term	2039	0.2	1.6	1.0	1.9	0.6
19	Long-term	2040	0.2	1.6	0.9	1.9	0.5
20	Long-term	2041	0.2	1.2	0.9	1.8	1.0
21	Long-term effects	2042	0.2	1.1	0.9	1.8	1.0
22	Long-term effects	2043	0.2	1.1	0.8	1.7	1.0
23	Long-term effects	2044	0.2	1.1	0.8	1.6	0.9
24	Long-term effects	2045	0.2	1.0	0.8	1.6	0.9
25	Long-term effects	2046	0.2	1.0	0.8	1.5	0.9
26	Long-term effects	2047	0.2	1.0	0.7	1.5	0.9
27	Long-term effects	2048	0.2	0.9	0.7	1.4	0.8
28	Long-term effects	2049	0.2	0.9	0.7	1.4	0.8

(continued)

Table 11.38 (continued)

SDAG							
t	Period	Year	Alternative 0	GoFooodLog	Safe and secure parking	ZLS	Intermodal terminal
29	Long-term effects	2050	0.2	0.9	0.7	1.3	0.8
30	Long-term effects	2051	0.2	0.9	0.6	1.3	0.8

Table 11.39 Performance of the alternatives across different time intervals for Friuli Venezia Giulia Region

Updated performance for Friuli Venezia Giulia Region					
Timeframe	Alternatives				
	Alternative 0	GoFooodLog	Safe and secure parking	ZLS	Intermodal terminal
2021–2024	2.747	4.290	3.311	5.312	2.087
2025–2031	3.871	13.925	7.788	19.755	0.720
2032–2041	2.321	17.385	10.528	21.216	6.577
2042–2051	2.069	9.911	7.464	15.146	8.797
2021–2051	11.0	45.5	29.1	61.4	18.2

Table 11.40 Performance of the alternatives across different time periods for Friuli Venezia Giulia Region

Updated performance for SDAG					
Timeframe periods	Alternatives				
	Alternative 0	GoFooodLog	Safe and secure parking	ZLS	Intermodal terminal
2021–2024	2.747	4.290	3.311	5.312	2.087
2021–2031	6.618	18.214	11.100	25.068	2.808
2021–2041	8.939	35.599	21.628	46.283	9.384
2021–2051	11.008	45.510	29.092	61.429	18.181

the preferred alternative for both SDAG and the Municipality of Gorizia, while the Friuli Venezia Giulia Region and the Port System Authority prefer the Simplified Logistics Zone (Fig. 11.34).

Considering the entire evaluation period 2021–2051, the preferred alternative for most stakeholders remains GoFoodLog, except for the Friuli Venezia Giulia Region, which prefers the Simplified Logistics Zone, a change that is noted only in the final interval (Table 11.47 and Fig. 11.35).

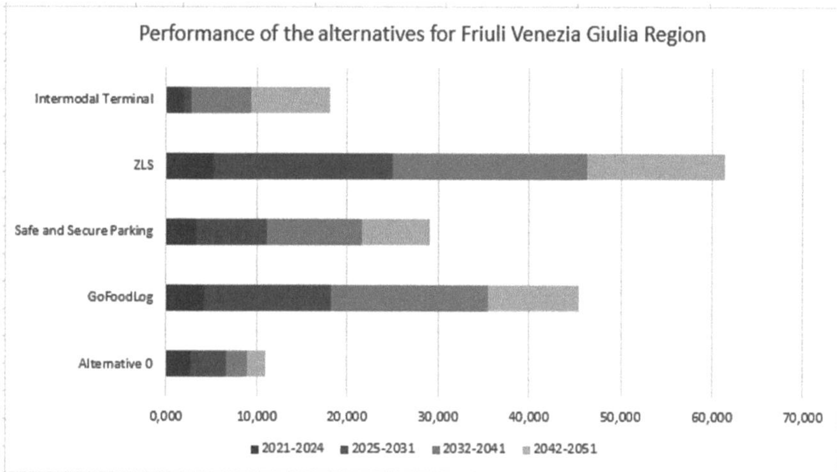

Fig. 11.27 Histogram of the performance of the alternatives in the different time intervals for Friuli Venezia Giulia Region, referring to Table 11.39

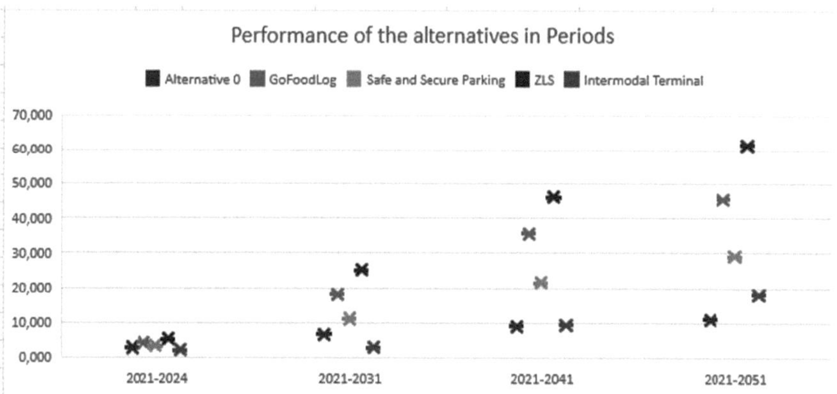

Fig. 11.28 Graph of comparison of the performance of the alternatives in the different time periods for Friuli Venezia Giulia Region, referring to Table 11.40

Table 11.41 Matrix of values of the alternatives in the time interval 2021-2024

Values of the alternatives in the time interval 2021–2024					
Stakeholder	Alternatives				
	Alternative 0	GoFooodLog	Safe and secure parking	ZLS	Intermodal terminal
SDAG	5.54	8.06	7.54	7.24	6.11
GO	0.00	2.91	0.45	2.31	0.00
PAS	4.17	6.31	4.42	5.81	3.17
FVG	2.75	4.29	3.31	5.31	2.09

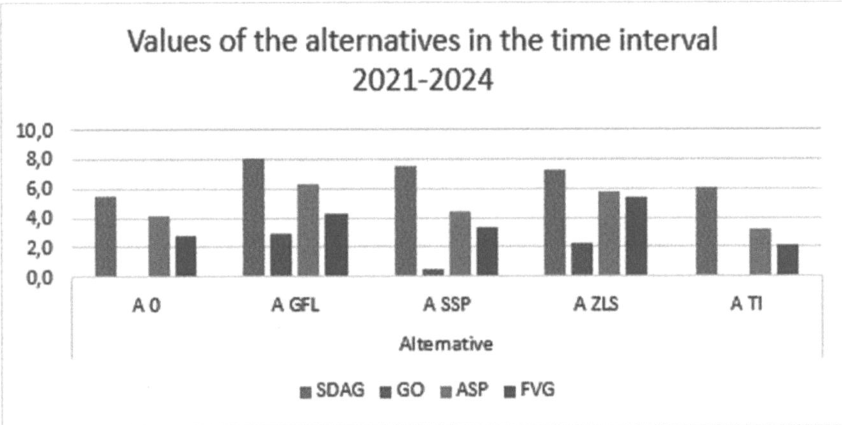

Fig. 11.29 Histogram of the values of the alternatives in the time interval 2021–2024, referring to Table 11.41

Table 11.42 Matrix of values of the alternatives in the time interval 2025–2031

Values of the alternatives in the time interval 2025–2031					
Stakeholder	Alternatives				
	Alternative 0	GoFooodLog	Safe and secure parking	ZLS	Intermodal terminal
SDAG	8.04	28.70	26.47	22.36	12.41
GO	0.00	18.57	2.76	15.29	0.79
PAS	5.60	19.41	7.75	16.78	0.72
FVG	3.87	13.92	7.79	19.76	0.72

It is evident that for all alternatives, SDAG has the highest performance as it is the most interested party, which was also indicated by the greater number of criteria expressed by SDAG compared to others. The performance of the alternatives for the Municipality of Gorizia is very heterogeneous, with high involvement but not direct. To integrate the evaluation into a single framework, it is necessary to consider the different involvement of the subjects, their freedom of action, and mutual influences. Therefore, the MACTOR Case Based method is applied.

11.6 The MACTOR Case-Based Method

This section refers to the methodology described in Chap. 9, "The MACTOR Method—Case-Based."

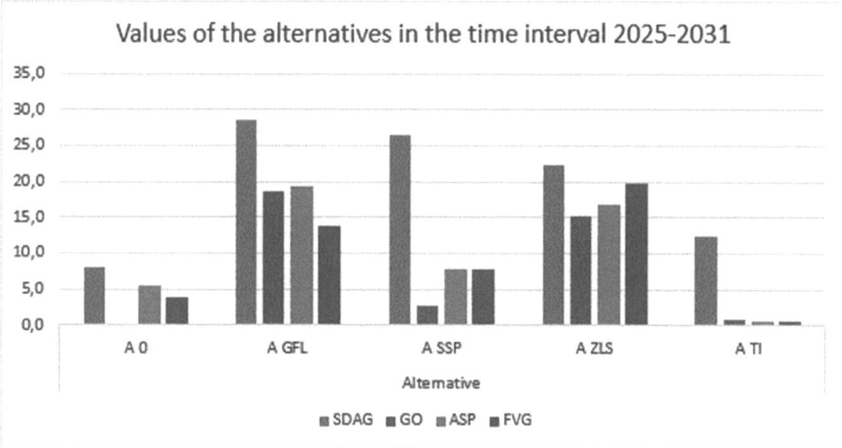

Fig. 11.30 Histogram of the values of the alternatives in the time interval 2025–2031, referring to Table 11.42

Table 11.43 Matrix of values of the alternatives in the time period 2021–2031

Values of the alternatives in the time period 2021–2031					
Stakeholder	Alternatives				
	Alternative 0	GoFooodLog	Safe and secure parking	ZLS	Intermodal terminal
SDAG	13.59	31.74	27.70	25.94	18.19
GO	0.00	21.48	3.21	17.60	0.79
PAS	9.77	25.71	12.16	22.59	3.90
FVG	6.62	18.21	11.10	25.07	2.81

11.6.1 Analysis Stakeholder Dynamics

The description of the stakeholders and their administrative configuration is detailed in Chap. 10, "Case Study: The Gorizia Freight Village".

An examination of the organizational structure of decision-makers and their respective governing bodies (Fig. 11.36—Schematic Diagram of Decision-Makers and Their Governing Bodies), their scope for action, and mutual influences (Fig. 61—Schematic Diagram of Direct and Indirect Influences among Stakeholders) reveals a clear interdependence among the actors.

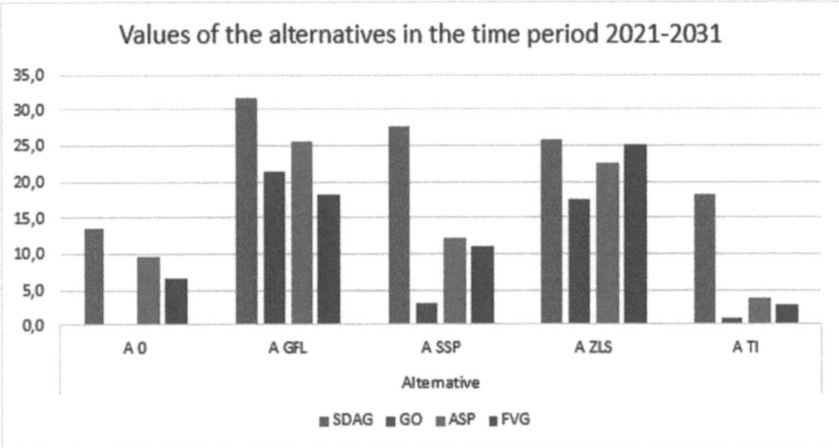

Fig. 11.31 Histogram of the values of the alternatives in the time period 2021–2031, referring to Table 11.43

Table 11.44 Matrix of values of the alternatives in the time interval 2032–2041

Values of the alternatives in the time interval 2032–2041					
Stakeholder	Alternatives				
	Alternative 0	GoFooodLog	Safe and secure parking	ZLS	Intermodal terminal
SDAG	8.60	26.55	21.14	22.28	17.49
GO	0.00	22.04	2.95	22.96	8.28
PAS	2.29	20.78	8.65	17.99	6.72
FVG	2.32	17.38	10.53	21.22	6.58

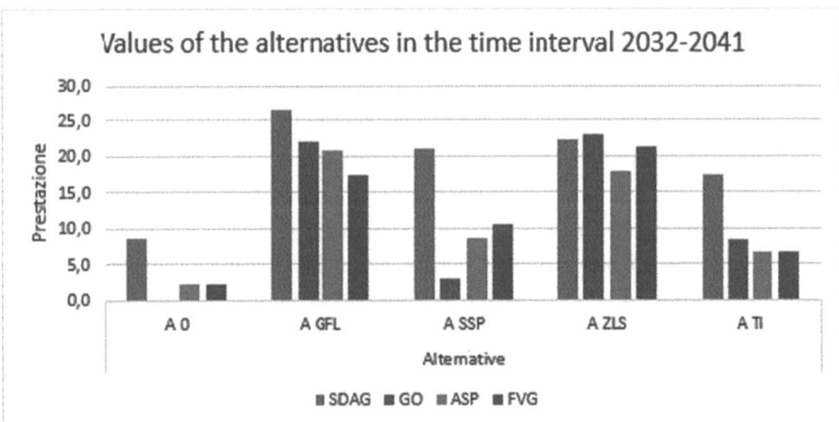

Fig. 11.32 Histogram of the values of the alternatives in the time interval 2032–2041, referring to Table 11.44

Table 11.45 Matrix of values of the alternatives in the time period 2021–2041

Values of the alternatives in the time period 2021–2041					
Stakeholder	Alternatives				
	Alternative 0	GoFooodLog	Safe and secure parking	ZLS	Intermodal terminal
SDAG	22.18	58.29	48.83	48.23	35.68
GO	0.00	43.53	6.16	40.56	9.07
PAS	12.05	46.49	20.81	40.59	10.61
FVG	8.94	35.60	21.63	46.28	9.38

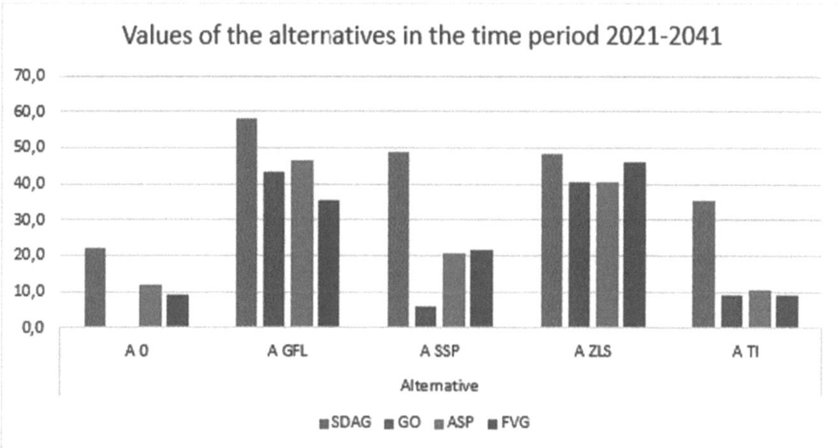

Fig. 11.33 Histogram of the values of the alternatives in the time period 2021–2041, referring to Table 11.45

Table 11.46 Matrix of values of the alternatives in the time interval 2042–2051

Values of the alternatives in the time interval 2042–2051					
Stakeholder	Alternatives				
	Alternative 0	GoFooodLog	Safe and secure parking	ZLS	Intermodal terminal
SDAG	6.10	15.29	15.09	15.36	14.31
GO	0.00	18.03	2.09	18.31	17.13
PAS	2.63	15.92	6.53	17.72	7.59
FVG	2.07	9.91	7.46	15.15	8.80

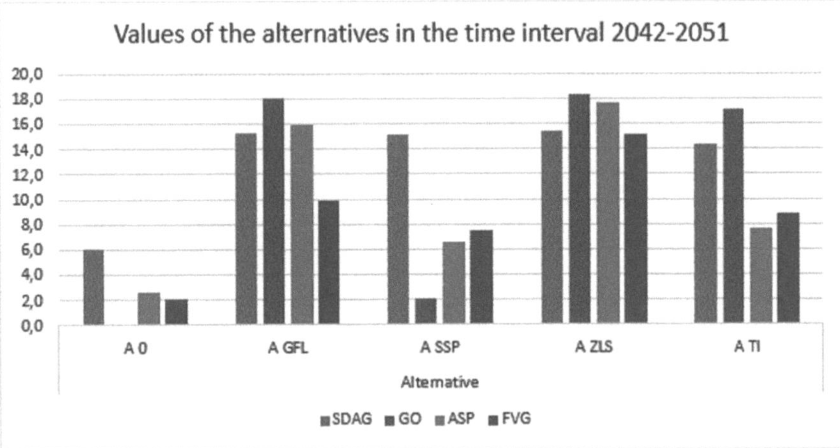

Fig. 11.34 Histogram of the values of the alternatives in the time interval 2042–2051, referring to Table 11.46

Table 11.47 Matrix of values of the alternatives in the time period 2021–2051

Values of the alternatives in the time period 2021–2051					
Stakeholder	Alternatives				
	Alternative 0	GoFooodLog	Safe and secure parking	ZLS	Intermodal terminal
SDAG	28.28	73.58	63.92	63.59	49.99
GO	0.00	61.56	8.25	58.88	26.20
PAS	14.69	62.41	27.35	58.31	18.20
FVG	11.01	45.51	29.09	61.43	18.18

11.6.1.1 Identification of High-Value Alternatives for Stakeholders

The alternatives considered are those that provide the best and worst performance for each stakeholder. This analysis is based on the 2021–2051time frame (see Table 11.47. Matrix of values of the alternatives in the time Period 2021–2051) (Table 11.48).

11.6.1.2 Identification of Strategic Issues

Strategic issues are explicitly outlined in the planning documents of each stakeholder. Specifically, the following documents are referenced:

- SDAG

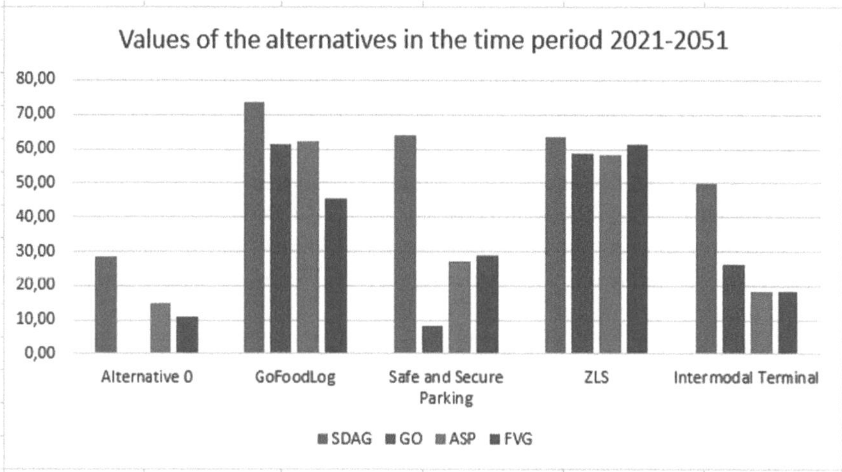

Fig. 11.35 Histogram of the values of the alternatives in the time period 2021–2051, referring to Table 11.47

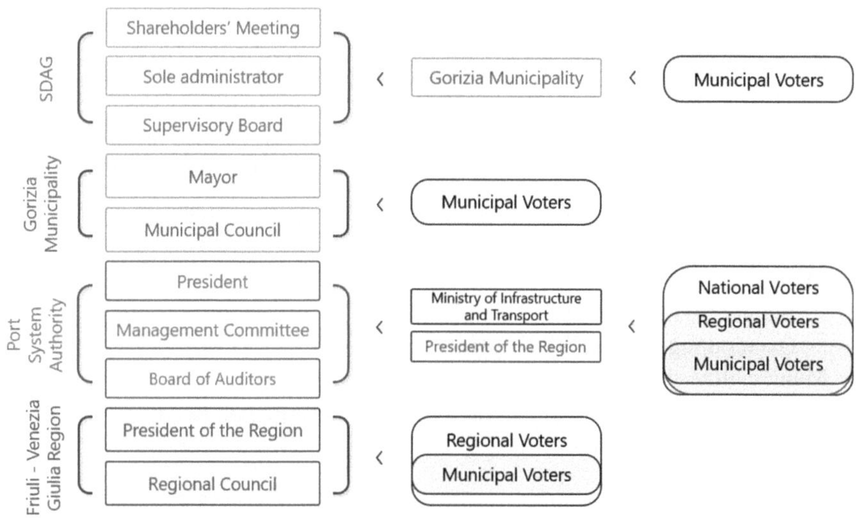

Fig. 11.36 Schematic diagram of decision-makers and their governing bodies

Table 11.48 Alternatives that maximize and minimize the evaluation

Alternatives that maximize and minimize the evaluation		
Stakeholder	Highest value alternative	Lowest value alternative
SDAG	GoFoodLog	Alternative 0
GO	GoFoodLog	Alternative 0
PSA	GoFoodLog	Alternative 0
FVG	ZLS	Alternative 0

"Documento programmatico pluriennale 2015–2022 Rev 02 ottobre 2018" (Multi-year Programmatic Document 2015–2022, Rev. 02, October 2018)

"Relazione sulla gestione del bilancio al 31/12/2017" (Financial Management Report as of 31/12/2017)

- Municipality of Gorizia

  "Piano Regolatore Generale del Comunale di Gorizia Vigente del 18/10//2001" (General Urban Plan (PRGC) in effect since 18/10/2001)

  "Studio socio-economico territoriale propedeutico alla revisione del Piano Regolatore Generale del Comunale di Gorizia" (Socio-economic and territorial study preparatory to the revision of the PRGC)

  "Revisione del P.R.G.C. vigente. Atto di indirizzo" (PRGC Revision: Policy Directive)
- Port System Authority

  "Piano Regolatore del Porto di Trieste" (Regulatory Plan of the Port of Trieste)
- Friuli Venezia Giulia Region

  "Piano Regionale delle Infrastrutture di Trasporto, della Mobilità delle Merci e della Logistica, L.R. 20 Agosto 2007, n 23, art ter e s.m.i." (Regional Plan for Transport Infrastructure, Freight Mobility, and Logistics, L.R. 20 August 2007, No. 23, Article ter and amendments.)

A comparative analysis of these documents and the criteria established by each stakeholder indicates a strong alignment between strategic issues and the criteria with the highest weight. Consequently, the analysis focuses on the two most significant criteria for each stakeholder, which collectively account for more than 50% of their decision-making weight.

The key criteria are as follows:

- SDAG

  Budget balance (28%)

  Overcoming marginality (25%)
- Municipality of Gorizia

  Economic and entrepreneurial development (49%)

  Community development (37%)
- Port System Authority

  System centrality and accessibility of inland terminals (55%)
- Friuli Venezia Giulia Region

  Territorial competitiveness (43%)

  Infrastructure rebalancing (29%)

To simplify the analysis, only the alternatives that maximize or minimize stakeholder performance concerning these strategic issues are considered (see Table 11.49—Diagram of Criteria with a Weight Greater than 50%). The corresponding table organizes stakeholders, strategic criteria, maximizing alternatives, and minimizing alternatives in a structured format (Table 11.50).

Table 11.49 Diagram of criteria with weight greater than 50%

Criteria identifying strategic issues				
Actor	Criteria	Acronym	Weight (%)	Total weight (%)
SDAG	Budget balance	C S1	29	54
	Overcoming marginality	C S3	25	
GO	Economic and entrepreneurial development	C G1	50	88
	Community development	C G2	38	
SPA	Centrality of the system and accessibility of inland ports	C A1	55	55
FVG	Regional territorial competitiveness	C R1	43	71
	Infrastructure rebalancing	C R3	29	

Table 11.50 Alternatives that maximize and minimize the performance of criteria identifying strategic issues

Alternatives that maximize and minimize the performance of criteria identifying strategic issues			
Actor	Criteria	Best performing alternatives by criterion	Worst performing alternatives by criterion
SADG	Budget balance Overcoming marginality Economic and entrepreneurial development Budget balance	Alternative 0	Alternative 0
		GoFoodLog	GoFoodLog
		Safe and secure parking	Safe and secure parking
		ZLS	ZLS
		Intermodal Terminal	Intermodal terminal
	Overcoming marginality	GoFoodLog	Alternative 0
		Safe and secure parking	
GO	Economic and entrepreneurial development	ZLS	Alternative 0
	Community development	GoFoodLog	Alternative 0
PSA	Centrality of the system and accessibility of inland ports	GoFoodLog	Alternative 0
FVG	Regional territorial competitiveness	ZLS	Alternative 0
	Infrastructure rebalancing	ZLS	Alternative 0

11.6.1.3 Identification of Priorities

During stakeholder workshops, each participant was asked to prioritize the alternatives. The resulting priority rankings are summarized in the accompanying table,

Table 11.51 Priority order of alternatives for each actor

Priority of alternatives for actors					
Actors	Alternatives				
	Alternative 0	GoFooodLog	Safe and secure parking	ZLS	Intermodal terminal
SDAG	5	1	2	4	4
GO	5	1	4	3	2
PAS	4	1	5	2	3
FVG	4	3	5	1	2

where the highest-priority alternatives are assigned a value of 1, with decreasing importance as the values increase (Table 11.51).

11.6.2 Stakeholder Positioning—Position Matrices

To understand stakeholder positioning relative to the value of alternatives, strategic issues, and priorities, position matrices are developed.

11.6.2.1 V-Matrix (Value Matrix)

Using Table 11.48, which lists alternatives that maximize and minimize performance, a graphical representation of stakeholder positions concerning the value of each alternative is constructed. The rules for compiling the V-Matrix, which assigns signs based on alternative values within the 2021–2051 timeframe (see Table 11.47), follow (Fig. 11.37).

The rules for compiling the V-Matrix, which assigns signs based on alternative values within the 2021–2051 timeframe (see Table 11.47), are the following (Table 11.52).

In this case Matrix V (Value positioning matrix with sign, Stakeholders × Alternatives) is (Table 11.53).

11.6.2.2 S-Matrix (Strategic Issue Matrix)

The S-Matrix represents stakeholder positioning concerning strategic issues. It is compiled using Table 11.50, which details the performance of alternatives in relation to strategic criteria, and the diagram in Fig. 11.38, which illustrates stakeholder positions regarding strategic issues. A positive sign indicates that an alternative maximizes a strategic criterion for a given stakeholder, while a negative sign indicates that it minimizes it.

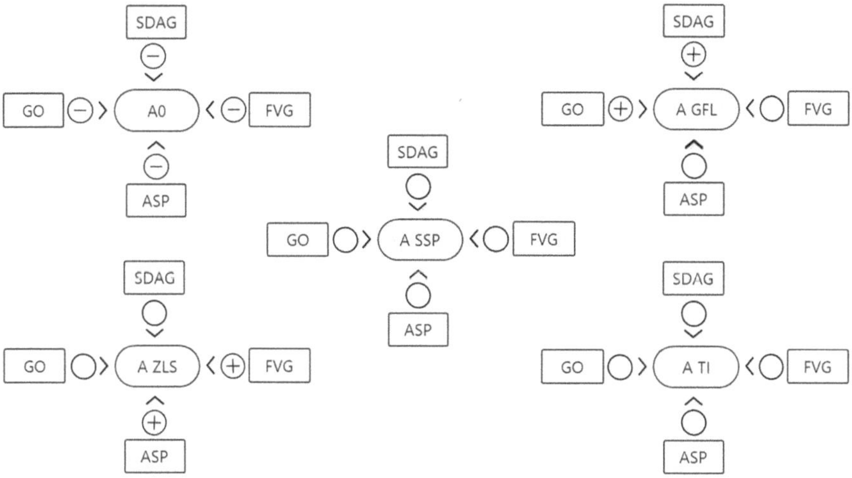

Fig. 11.37 Actors' positions relative to the alternatives with the best values

Table 11.52 Compilation rules of the V matrix

Compilation rules of the V matrix	
+1	The alternative has the best performance for the stakeholder
0	The alternative has intermediate performance for the stakeholder
−1	The alternative has the worst performance for the stakeholder

Table 11.53 Value positioning matrix with sign, stakeholders × alternatives

V matrix of Stakeholders' positioning alternatives for the values of the alternatives in time period 2021–2051

Stakeholder	Alternatives					Tot+	Tot−	Tot
	A 0	A GFL	A SSP	A ZLS	A TI			
SDAG	−1	1	0	0	0	1	−1	0
GO	−1	1	0	0	0	1	−1	0
PSA	−1	1	0	0	0	1	−1	0
FVG	−1	0	0	1	0	1	−1	0
Tot+	0	3	0	1	0			
Tot−	−4	0	0	0	0			
Tot	−4	3	0	1	0			

The S matrix of positioning with respect to strategic issues (Table 11.55—S matrix of positioning with respect to strategic issues) can therefore be completed according to the rules defined in Chap. 9, which are provided here (Table 11.54).

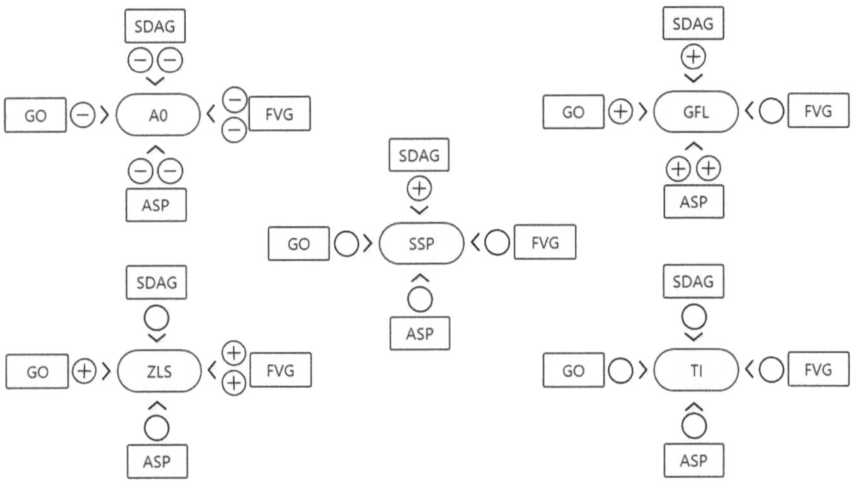

Fig. 11.38 Actors' positions relative to the alternatives with the best values for strategic issues

Table 11.54 Compilation rules of the S matrix

Compilation rules of the V matrix	
+2	The alternative maximizes both criteria representing strategic issues for the stakeholder
+1	The alternative maximizes one criterion representing strategic issues for the stakeholder
0	The alternative has intermediate performance for the stakeholder
−1	The alternative minimizes one criterion representing strategic issues for the stakeholder
−2	The alternative minimizes both criteria representing strategic issues for the stakeholder

Table 11.55 S matrix of positioning with respect to strategic issues

S matrix of positioning with respect to strategic issues in time period 2021–2051								
Stakeholder	Alternatives					Tot+	Tot−	Tot
	A 0	A GFL	A SSP	A ZLS	A TI			
SDAG	−1	1	1	0	0	2	−1	1
GO	−2	1	0	1	0	2	−2	0
PSA	−2	2	0	0	0	2	−2	0
FVG	−2	0	0	2	0	2	−2	0
Tot+	0	4	1	3	0			
Tot−	−7	0	0	0	0			
Tot	−7	4	1	3	0			

In this case Matrix S (Value positioning matrix with sign, Stakeholders × Alternatives) is.

The S-Matrix reveals clear strategic preferences: ASP and FVG exhibit strong alignment with the GoFoodLog and ZLS alternatives, respectively, while SDAG and the Municipality of Gorizia have more diversified interests. Additionally, there is a broad consensus that Alternative 0 is the least desirable. Further analysis of the matrix suggests that, while most stakeholders align on certain alternatives, FVG consistently diverges, showing a distinct preference for ZLS. SDAG, on the other hand, demonstrates broader interest distribution, whereas the Friuli Venezia Giulia Region remains focused on just two alternatives.

11.6.2.3 P-Matrix (Priority Matrix)

As detailed in Chap. 9, the priority rankings from previous tables are transformed into a hierarchical scale ranging from 2 to −2, reflecting stakeholder interest levels in each alternative. The resulting priority positioning matrix, P, is then analysed for areas of convergence and divergence (Table 11.56).

11.6.3 Defining Convergence and Divergence

To assess convergence and divergence among stakeholders, two matrices are constructed:

- C-Matrix (Actor-to-Actor Convergence Matrix)
- D-Matrix (Actor-to-Actor Divergence Matrix).

These matrices are derived from pairwise comparisons of values from the V, S, and P matrices to determine the extent of alignment or misalignment in stakeholder preferences, strategic issues, and priorities.

For illustrative purposes, Table 11.57 outlines the calculation of convergence and divergence between SDAG and the Municipality of Gorizia concerning strategic issues comparing the first two rows of the S matrix.

Table 11.56 P matrix of positioning the priority of alternatives for the actors

P matrix of actors					
Actors	Alternatives				
	Alternative 0	GoFooodLog	Safe and secure parking	ZLS	Intermodal terminal
SDAG	−2	2	1	−1	−1
GO	−2	2	−1	0	1
PAS	−1	2	−2	1	0
FVG	−1	0	−2	2	1

Table 11.57 Calculation of convergence and divergence between SDAG and the Municipality of Gorizia with respect to strategic issues based on the S matrix

S matrix of positioning with respect to strategic issues in time period 2021–2051						
Stakeholder	Alternatives					Tot
	A 0	A GFL	A SSP	A ZLS	A TI	
SDAG	−1	1	1	0	0	
GO	−2	1	0	1	0	
PSA	−2	2	0	0	0	
FVG	−2	0	0	2	0	
Conv	0.5	1	0.5	0.5	1	3.5
Div	−0.5	0	−0.5	−0.5	0	−1.5

Table 11.58 Cs matrix of actors' convergence with respect to strategic issue

Cs convergence matrix strategic issue				
Stakeholders	SDAG	GO	PSA	FVG
SDAG	–	3.5	3.5	3
GO	3.5	–	4	4
PSA	3.5	4	–	4
FVG	3	4	4	–

Table 11.59 Ds matrix of actors' divergence with respect to strategic issues

Ds divergence matrix strategic issue				
Stakeholders	SDAG	GO	PSA	FVG
SDAG	–	−1.5	−1.5	−2
GO	−1.5	–	1	1
PSA	−1.5	−1	–	−1
FVG	−2	−1	−1	–

The analysis shows partial convergence for Alternatives Zero, Safe and Secure Park, and Simplified Logistics Zone (each weighted at 0.5), whereas full convergence is found for GoFoodLog and the Intermodal Terminal (each receiving a weight of 1). The total convergence score is 3.5, while the divergence score is −1.5, indicating weak divergence without a total misalignment.

The procedure is then applied to all stakeholder pairs, with results summarized in Tables 11.58 (Strategic Issue Convergence Matrix) and 59 (Strategic Issue Divergence Matrix). Corresponding convergence and divergence diagrams visually represent the findings, with thicker lines indicating stronger alignment (Table 11.59).

The first diagrams of convergence and divergence can thus be constructed; the thicker the line, the greater the convergence between the actors (Fig. 11.39).

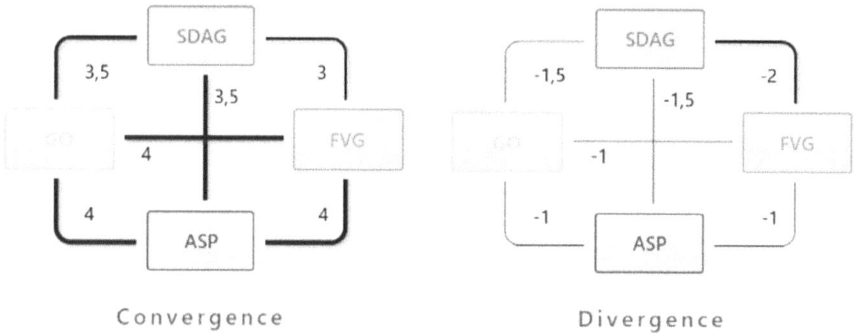

Fig. 11.39 Diagrams of convergence and divergence with respect to strategic issues

The diagrams highlight a strong consensus on alternatives that maximize strategic issues, with minimal divergence. The most significant divergence occurs between SDAG and FVG, while the Municipality of Gorizia and ASP exhibit the highest convergence.

Similarly, additional convergence (Cv) and divergence (Dv) matrices are constructed based on the V-Matrix values (Tables 11.60 and 11.61).

The first diagrams of convergence and divergence can thus be constructed; the thicker the line, the greater the convergence between the actors (Fig. 11.40).

Comparing the results reveals that FVG remains isolated in alternative valuation, while other stakeholders exhibit strong mutual alignment.

Table 11.60 Cv matrix of actors' convergence with respect to the values of the alternatives over the period 2021–2051

Cv convergence matrix values of alternative				
Stakeholders	SDAG	GO	PSA	FVG
SDAG	–	5	5	3
GO	5	–	5	3
PSA	5	5	–	3
FVG	3	3	3	–

Table 11.61 Dv matrix of actors' divergence with respect to the values of the alternatives over the period 2021–2051

Dv divergence matrix values of alternative				
Stakeholders	SDAG	GO	PSA	FVG
SDAG	–	0	0	−2
GO	0	–	0	−2
PSA	0	0	–	−2
FVG	−2	−2	−2	–

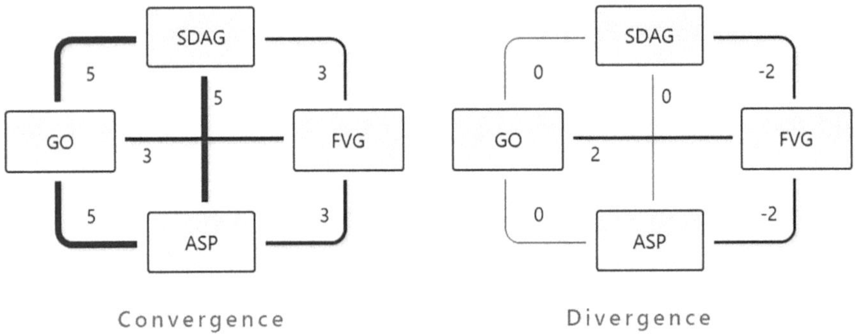

Fig. 11.40 Diagrams of convergence and divergence with respect to the values of the alternatives over the period 2021–2051

A notable shift occurs in the transition from the S-Matrix to the V-Matrix, where strategic issue alignment brings together the Municipality of Gorizia, ASP, and FVG, whereas in the value analysis, SDAG aligns more closely with Gorizia and ASP.

Similarly, additional convergence (Cv) and divergence (Dv) matrices are constructed based on the P-Matrix values (Tables 11.62 and 11.63).

The first diagrams of convergence and divergence can thus be constructed; the thicker the line, the greater the convergence between the actors (Fig. 11.41).

The greatest convergence with respect to the priorities of the alternatives is evident between ASP and FVG, while the greatest divergence is between SDAG and FVG. Therefore, there is a further shift in positions compared to those emerging from the S and V matrices, although the distance between SDAG and FVG remains.

Table 11.62 Cp matrix of actors' convergence with respect to stakeholders' priorities

Cp convergence matrix priorities of alternative				
Stakeholders	SDAG	GO	PSA	FVG
SDAG	–	2.5	2	1
GO	2.5	–	3	3
PSA	2	3	–	3.5
FVG	1	3	3.5	–

Table 11.63 Dp matrix of actors' divergence with respect to stakeholders' priorities

Dp divergence matrix priorities of alternative				
Stakeholders	SDAG	GO	PSA	FVG
SDAG	–	−2.5	−3	−4
GO	−2.5	–	−2	−2
PSA	−3	−2	–	−1.5
FVG	−4	−2	−1.5	–

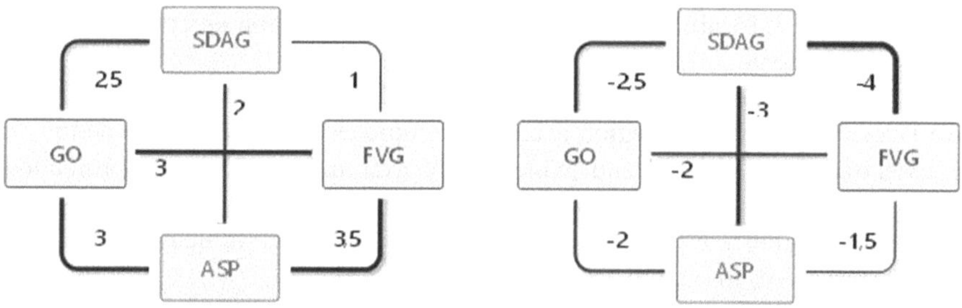

Fig. 11.41 Diagrams of convergence and divergence with respect to stakeholders' priorities

These initial matrices, while providing valuable insights, remain somewhat simplistic as they only account for numerical alignments. To refine the model and better approximate real-world dynamics, it is essential to introduce an additional dimension: the power relations among stakeholders.

11.6.4 Analysis of Power Dynamics

The interaction between potential convergences and divergences is not solely determined by individual actor preferences, but also by their capacity to assert influence over others. While significant divergences exist among actors regarding various alternatives, simultaneous convergence points emerge. Critically, power relations and interdependencies between actors play a pivotal role in shaping these complex dynamics.

11.6.4.1 Institutional Actors and Their Interactions

SDAG (Stazioni Doganali Autoportuali di Gorizia). SDAG serves as the direct administrative entity managing the interport. However, its governance structure is fundamentally influenced by the Municipality of Gorizia. Specifically, under Article 2449 of the Civil Code, the Municipality appoints the Sole Administrator or, in the case of a Board of Directors, nominates the majority of its members.

Gorizia Municipality. The electorate of Gorizia Municipality elects the Mayor and Municipal Council, who are responsible for approving the General Regulatory Plan. This plan designates the Sant'Andrea area as a regional merchandise interchange zone (Autoporto), classified as Homogeneous Zone H1, in alignment with regional planning objectives.

Port System Authority of the Eastern Adriatic. This authority encompasses multifaceted responsibilities including strategic direction, programming, coordination, promotion, and oversight of port operations. While possessing administrative autonomy, the President's appointment is a collaborative process involving the Minister of Infrastructure and Transport, contingent upon agreement with the Regional President.

Autonomous Region of Friuli Venezia Giulia. In accordance with Legislative Decree 111/2004, the region exercises comprehensive regulatory functions in regional transportation. Its mandate includes programming, guidance, coordination, monitoring, and supervisory roles. Furthermore, pursuant to Article 4 of its Special Statute, the region governs urban planning and territorial strategy, establishing strategic objectives through the Regional Territorial Plan.

11.6.4.2 Power Dynamics and Influence Matrix

To clarify the aforementioned relationships, the following diagrams and subsequent influence matrices have been developed (Fig. 11.42).

The influence matrix, compiled according to the methodology presented in Chap. 9, where C_i indicates the independence coefficient and C_d is the dependence coefficient, is as follows (Tables 11.64 and 11.65).

A detailed influence matrix reveals nuanced power relationships among these institutional actors. The Port System Authority emerges as the most influential entity, wielding significant indirect influence over both SDAG and the Friuli Venezia Giulia Region. Conversely, SDAG occupies the most vulnerable position, despite being the direct administrator of the Sant'Andrea interport, primarily due to its dependency on the Gorizia Municipality.

This structural configuration potentially compels SDAG to seek strategic alliances, particularly with the Port System Authority, to advance its institutional objectives.

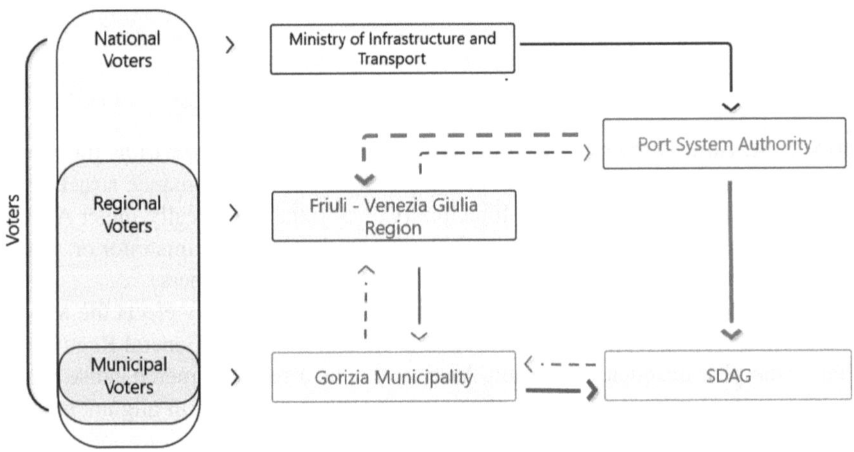

Fig. 11.42 Schematic diagram of direct and indirect influences between actors. A solid arrow represents a direct influence, while a dashed arrow represents an indirect influence

Table 11.64 Influence matrix

Influence matrix					
Stakeholders	SDAG	GO	PSA	FVG	C_i
SDAG	–	1	0	0	1
GO	4	–	0	1	5
PSA	3	0	–	3	6
FVG	1	2	2	–	5
C_d	8	3	2	4	17

Table 11.65 Rules for completing the I matrix

0	No influence
1	Weak indirect influence
2	Weak Direct influence
3	Strong indirect influence
4	Strong direct influence

Table 11.66 Independence coefficients, dependence coefficients, relative independence, relative dependence, power balance

Influence weighting coefficients					
	C_i	C_d	C_{ir}	C_{idr}	C*
SDAG	1	8	0.06	0.01	1.62
GO	5	3	0.29	0.18	8.08
PAS	6	2	0.35	0.26	9.70
FVG	5	4	0.29	0.16	8.08

Starting from the Influence matrix, the independence and relative dependence coefficients for all actors are obtained, through which c is derived, which is the power balance coefficient, through the procedure described in Chap. 9 (Table 11.66 and Fig. 11.43).

11.6.5 Stakeholder Positioning Matrices Balanced for Influence

The V matrix of the values of the alternatives for the actors with the power balance vector will result in the Vi matrix as follows (Table 11.67).

Similarly, the Si matrix of positioning strategic issues will be weighted by power relations (Table 11.68).

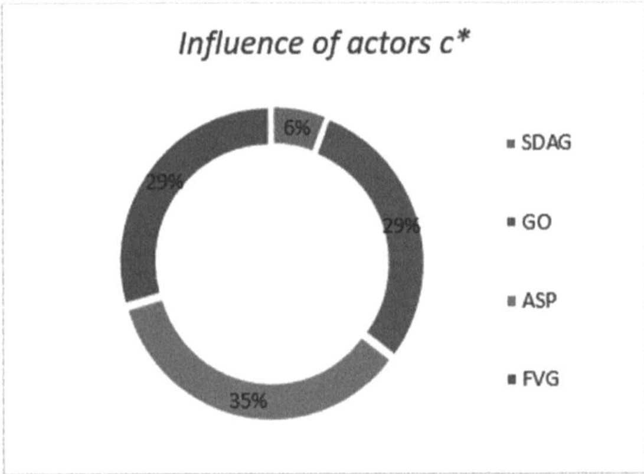

Fig. 11.43 Explanatory chart of actors' influence, referring to Table 11.66

Table 11.67 V_i matrix of positioning the values of alternatives weighted by influence

Vi-matrix					
Stakeholder	Alternatives				
	A 0	A GFL	A SSP	A ZLS	A TI
SDAG	−1.6	1.6	0.0	0.0	0.0
GO	−8.1	8.1	0.0	0.0	0.0
PSA	−9.7	9.7	0.0	0.0	0.0
FVG	−8.1	0.0	0.0	8.1	0.0

Table 11.68 SI matrix of positioning strategic issues weighted by influence

Si-matrix					
Stakeholder	Alternatives				
	A 0	A GFL	A SSP	A ZLS	A TI
SDAG	−1.6	1.6	1.6	0.0	0.0
GO	−16.2	8.1	0.0	8.1	0.0
PSA	−19.4	19.4	0.0	0.0	0.0
FVG	−16.2	0.0	0.0	16.2	0.0

Similarly, the P matrix of positioning priority for actors will be weighted by the power balancing vector, resulting in the Pi matrix (Table 11.69).

Table 11.69 PI matrix of priority weighted by influence

Pi-matrix					
Stakeholder	Alternatives				
	A 0	A GFL	A SSP	A ZLS	A TI
SDAG	−3.23	3.23	1.62	−1.62	−1.62
GO	−16.17	16.17	−8.08	0.00	8.08
PSA	−9.70	19.40	−19.40	9.70	0.00
FVG	−8.08	0.00	−16.17	16.17	8.08

11.6.6 Definition of Convergence and Divergence Balanced by Influence

At this stage, it is possible to define the new convergence and divergence matrices, adjusted according to power dynamics. This process involves a pairwise comparison of actors. The procedure is consistent with that used for previous matrices.

The pairwise comparison is achieved by analyzing the values of Matrix Si, Vi, Pi, then is calculated the difference between each pair of values.

These differences allow for the determination of distances between positions within each pair of actors. The smaller the distance, the higher the convergence on a given alternative; conversely, a greater distance indicates a higher level of divergence. The distances are then transformed into a convergence-divergence scale while maintaining proportionality among the intervals.

Following this transformation, convergence and divergence values can be calculated for each alternative.

The aggregate convergence values are obtained by summing those of each pair, applying the same procedure to calculate divergence. The results are presented in the Tables 11.70 and 11.71.

Subsequently, the first diagrams illustrating convergence and divergence can be constructed. The thickness of the lines indicates the degree of convergence between actors.

Table 11.70 Csi matrix of convergence for strategic issues adjusted for power relations (actors × actors)

Csi convergence matrix strategic issues adjusted for power relations				
Stakeholders	SDAG	GO	PSA	FVG
SDAG	0.0	17.3	15.7	16.5
GO	17.3	25.0	19.3	21.0
PSA	15.7	19.3	25.0	15.3
FVG	16.5	21.0	15.3	25.0

Table 11.71 Dsi matrix of divergence for strategic issues adjusted for power relations (actors × actors)

Dsi divergence matrix strategic issues adjusted for power relations				
Stakeholders	SDAG	GO	PSA	FVG
SDAG	0.0	7.7	9.3	8.5
GO	7.7	0.0	5.7	4.0
PSA	9.3	5.7	0.0	9.7
FVG	8.5	4.0	9.7	0.0

Examining the most significant aspects of the diagrams in Fig. 11.44 and comparing them with those in Fig. 11.39 reveals the crucial role of mutual influences in actor dynamics. Notably, the alignment among the Municipality of Gorizia, the Port System Authority (ASP), and the Friuli Venezia Giulia Region (FVG) has strengthened, despite increased divergence between the Region and ASP. This shift has brought the Municipality of Gorizia closer to the Friuli Venezia Giulia Region. Regarding divergence, the Municipality of Gorizia exhibits the least divergence compared to the other actors.

Similarly, starting from the Vi matrix (Table 11.72), the convergence and divergence matrices for the values of alternatives, adjusted for influence, can be derived using the same steps (Table 11.73).

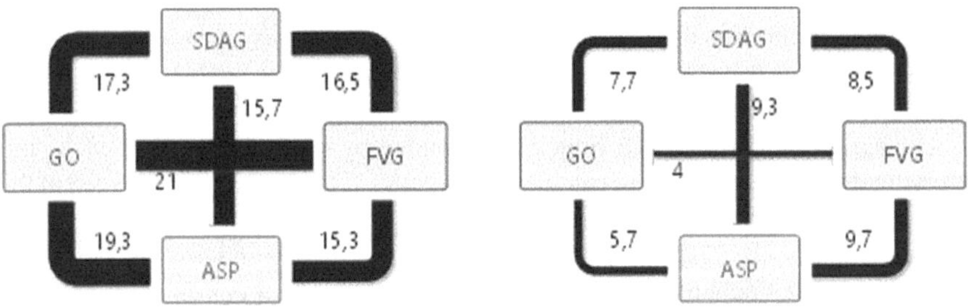

Fig. 11.44 Diagrams of convergence and divergence on strategic issues adjusted for power relations

Table 11.72 Cvi matrix of convergence on alternative values adjusted for influence

Cvi convergence matrix alternative values adjusted for power relations				
Stakeholders	SDAG	GO	PSA	FVG
SDAG	25.0	16.9	14.9	14.9
GO	16.9	25.0	23.0	14.9
PSA	14.9	23.0	25.0	12.9
FVG	14.9	14.9	12.9	25.0

Table 11.73 Dvi matrix of divergence on alternative values adjusted for influence

Dvi divergence matrix alternative values adjusted for power relations				
Stakeholders	SDAG	GO	PSA	FVG
SDAG	0.0	8.1	10.1	10.1
GO	8.1	0.0	2.0	10.1
PSA	10.1	2.0	0.0	12.1
FVG	10.1	10.1	12.1	0.0

Subsequently, the first diagrams illustrating convergence and divergence can be constructed. The thickness of the lines indicates the degree of convergence between actors.

The results highlight that the greatest convergence regarding alternative values, adjusted for power relations, occurs between the Municipality of Gorizia and ASP, whereas the highest divergence is observed between ASP and FVG.

Comparing the diagrams in Fig. 11.45 with those in Fig. 11.40, it becomes evident that the introduction of power relations has reduced the convergence between the Municipality and SDAG, which is now also in conflict with ASP without moving closer to the Friuli Venezia Giulia Region. Additionally, all divergence values have increased.

Following a similar approach, starting from the Pi matrix (Table 11.74), convergence and divergence matrices for alternative priorities can be obtained using the same steps. Therefore, only the results are presented here (Table 11.75).

Subsequently, the first diagrams illustrating convergence and divergence can be constructed. The thickness of the lines indicates the degree of convergence between actors (Fig. 11.46).

In this case, the Port System Authority exhibits convergence with both the Friuli Venezia Giulia Region and the Municipality of Gorizia, despite the latter two not being aligned with each other. The Municipality of Gorizia's second-highest convergence is with SDAG. Regarding divergence, there is a notable distance between SDAG and both the Port System Authority and the Friuli Venezia Giulia Region.

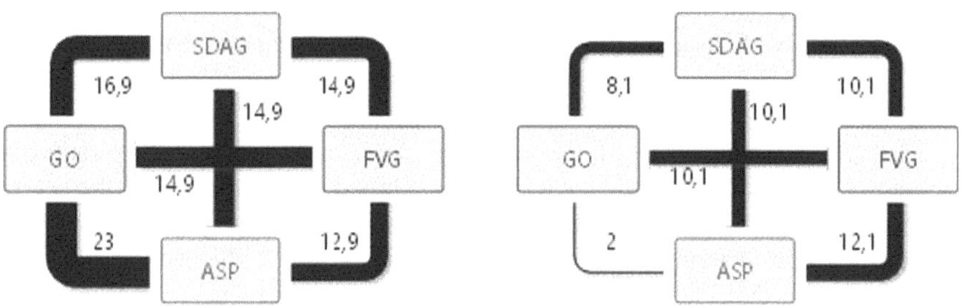

Fig. 11.45 Diagrams of convergence and divergence on alternative values adjusted for power relations

Table 11.74 Cpi matrix of convergence on priorities adjusted for power relations (actors × actors)

Cpi convergence matrix priorities adjusted for power relations				
Stakeholders	SDAG	GO	PSA	FVG
SDAG	25.0	13.3	10.9	11.7
GO	13.3	25.0	15.3	12.9
PSA	10.9	15.3	25.0	15.3
FVG	11.7	12.9	15.3	25.0

Table 11.75 Dpi matrix of divergence on priorities adjusted for power relations (actors × actors)

Dpi divergence matrix priorities adjusted for power relations				
Stakeholders	SDAG	GO	PSA	FVG
SDAG	0.0	11.7	14.1	13.3
GO	11.7	0.0	9.7	12.1
PSA	14.1	9.7	0.0	9.7
FVG	13.3	12.1	9.7	0.0

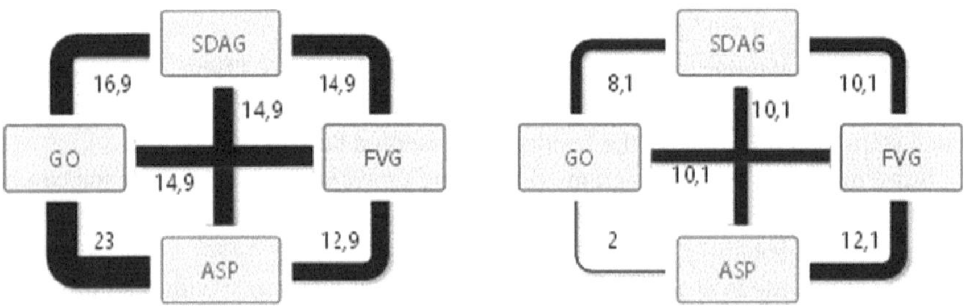

Fig. 11.46 Diagrams of convergence and divergence on alternative priorities weighted by power relations

Comparing these diagrams with those in Fig. 11.41, which depict alternative priority convergence and divergence without considering power relations, a shift in the Municipality of Gorizia's position is evident. The municipality has moved closer to SDAG while distancing itself from FVG, though the separation between SDAG and FVG remains unchanged.

Through the integration of the collected data, a comprehensive perspective emerges, confirming the challenge of achieving broad consensus among the actors. Weighting the results based on influence has made the fragmentation of the system more explicit, as demonstrated by the breakdown of major alliances. However, it is worth noting that FVG exhibits the highest overall divergence, while ASP has achieved the highest average convergence. Nevertheless, ASP's influence remains insufficient to exert a decisive effect on other actors.

Table 11.76 Matrix D, decision matrix

Decision matrix					
Stakeholder	Alternatives				
	Alternative 0	GoFooodLog	Safe and secure parking	ZLS	Intermodal terminal
SDAG	−6.47	6.47	3.23	−1.62	−1.62
GO	−40.42	32.34	−8.08	8.08	8.08
PAS	−38.81	48.51	−19.40	9.70	0.00
FVG	−32.34	0.00	−16.17	40.42	8.08

Should FVG and ASP form an alliance, a central point of convergence could emerge, potentially attracting the alignment of other actors within the system.

11.7 The Decision Matrix

The comparative evaluation of previously obtained values is conducted through the elaboration of Decision Matrix D, as described in Chap. 9. This matrix expressively captures the potential for realizing specific alternatives across different actors, taking into account their declared preferences, strategic considerations, and priorities. Crucially, the realization potential integrates each actor's scope of action, carefully balanced against prevailing power dynamics (Table 11.76).

The subsequent four graphical representations visually articulate each actor's individual capacity to realize the proposed alternatives. When an alternative's bar extends into positive values, it signifies significant interest for the given actor, with the bar's length corresponding directly to enhanced realization potential. Conversely, bars extending into negative values indicate minimal actor interest, with increased negative extension proportionally diminishing implementation prospects (Fig. 11.47).

For SDAG, GoFoodLog emerges as the most independently viable alternative, with Safe and Secure Parking representing a secondary option that can be autonomously implemented, albeit with notably reduced preference. The Terminal Intermodal and ZLS alternatives prove challenging to realize without broader stakeholder consensus, while Alternative 0 demonstrates no practical implementation potential, being the least attractive option for all involved actors (Fig. 11.48).

From the Municipality of Gorizia's perspective, GoFoodLog stands as the most individually feasible alternative. The Terminal Intermodal and ZLS alternatives both present interesting prospects, positioning them positively on the graphical representation. The municipality might actively seek collaborative agreements to advance these options. Conversely, the Safe and Secure Parking alternative holds minimal appeal, with Alternative 0 again demonstrating no realistic implementation potential (Fig. 11.49).

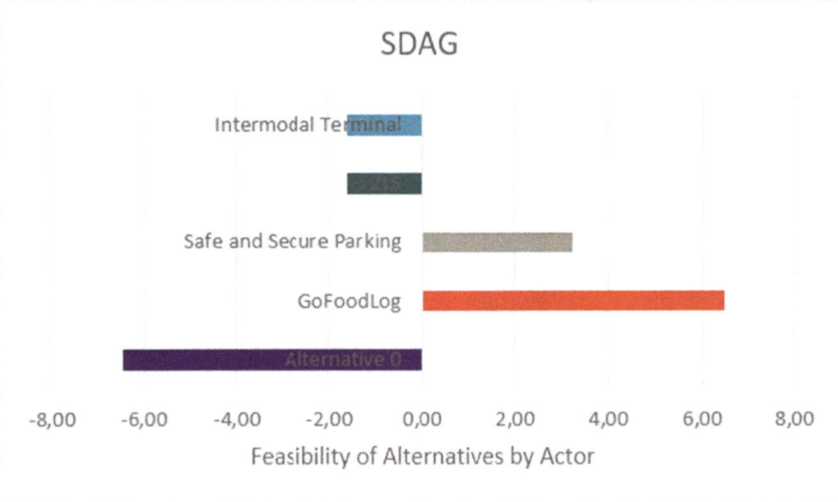

Fig. 11.47 Graphical representation of SDAG's potential to independently implement alternatives

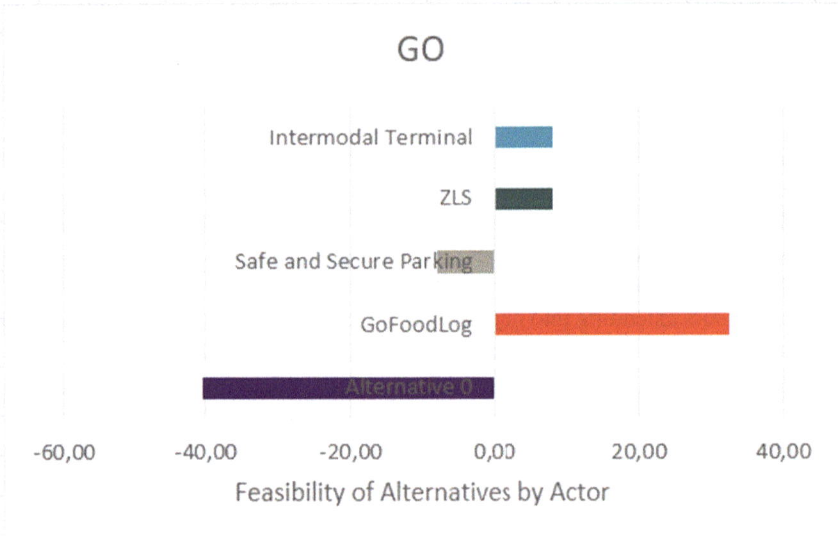

Fig. 11.48 Graphical representation of municipality of Gorizia's potential to independently implement alternatives

For the Port System Authority, GoFoodLog represents the most readily implementable alternative, with ZLS emerging as a close second. The authority will likely pursue active collaboration to realize this alternative, whereas Safe and Secure Parking receives minimal consideration. The Terminal Intermodal alternative may

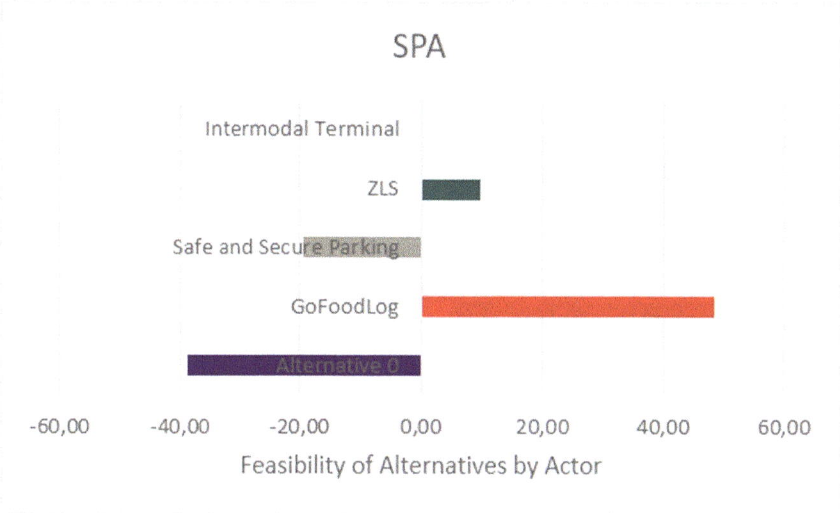

Fig. 11.49 Graphical representation of Port System Authority's potential to independently implement alternatives

serve as a strategic negotiation tool, positioned neutrally within their strategic framework (Fig. 11.50).

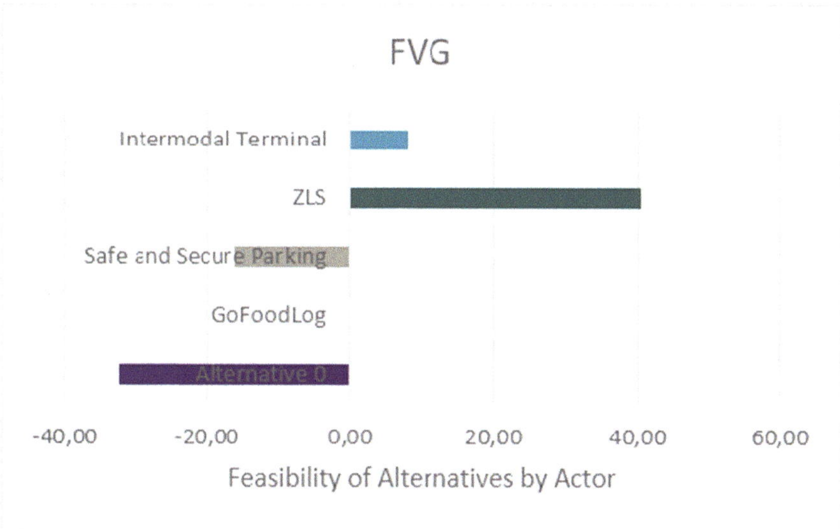

Fig. 11.50 Graphical representation of Friuli Venezia Giulia Region's potential to independently implement alternatives

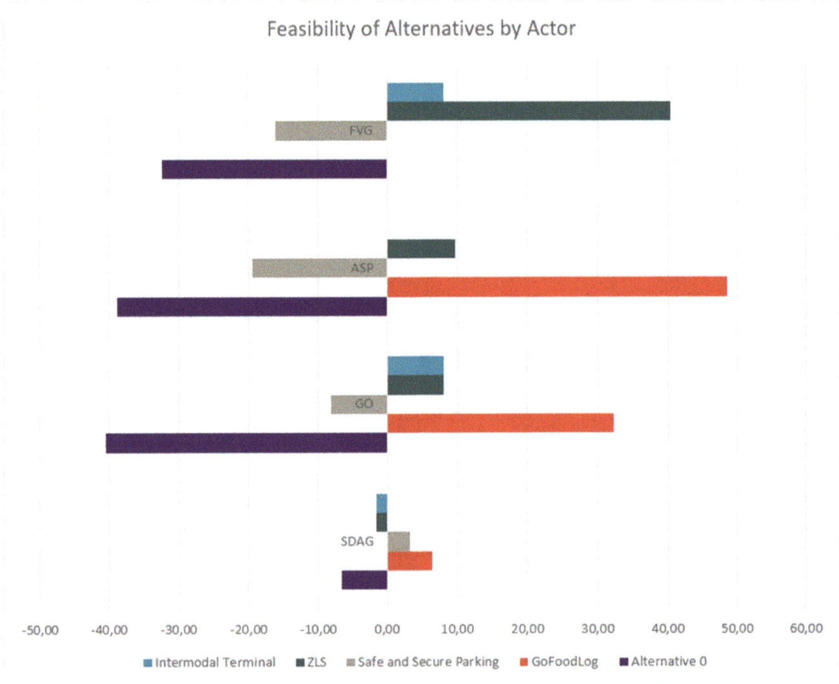

Fig. 11.51 Graphical representation of each actor's potential to independently implement alternatives

The Friuli Venezia Giulia Region identifies ZLS as its most independently achievable alternative. The region may proactively seek collaborative agreements to implement the Terminal Intermodal project, potentially leveraging GoFoodLog as a negotiation instrument, given its perceived neutrality. The Safe and Secure Parking alternative remains unlikely to garner significant regional interest, with Alternative 0 continuing to lack implementation viability.

11.7.1 Critical Interpretative Observations

The subsequent graphical synthesis illuminates each actor's alternative realization potential. Notably, the Port System Authority and Friuli Venezia Giulia Region demonstrate substantially higher capacities for individual alternative implementation compared to SDAG. The Municipality of Gorizia occupies an intermediary position, crucially facilitating the transcendence of previously identified systemic fragmentation (Fig. 11.51).

The Gorizia logistics hub encounters significant challenges, constrained by a robust Friulian industrial hinterland and the gravitational economic pull of the regional capital, which concentrates investment opportunities.

The GoFoodLog alternative emerges consistently as the most compelling option across evaluation phases. Its transformative potential for SDAG positions it as a pivotal reference point within the Friuli Venezia Giulia agri-food logistics landscape. The proposed spatial reconfiguration and establishment of a novel cold storage infrastructure within the regional logistics system promise advanced product management solutions, benefiting the entire supply chain—a prospect recognized by both the Gorizia Municipality and Port System Authority.

This alternative harmonizes seamlessly with the Friuli Venezia Giulia Region's preferred Logistic Simplified Zone approach, which aims to cultivate a facilitative economic regime supporting logistics and port-related economic activities. This strategic orientation aligns with the Port System Authority's objectives, focusing on entrepreneurial settlement and investment project valorisation. The approach generates meaningful consensus among stakeholders, presenting a potential pathway beyond current marginal economic conditions.

The analysis reveals that within a polycentric yet politically and territorially cohesive system like Friuli Venezia Giulia, regional revitalization necessitates developing novel alliance structures that transform competitive dynamics into collaborative frameworks.

In conclusion, for SDAG to emerge as a strategic intermodal platform facilitating Mitteleuropean exchanges, it must functionally and operationally integrate with the regional system. Simultaneously, the regional ecosystem must promote an organizational model that networks and shares services.

Leveraging its transborder positioning between Italy and Slovenia, SDAG could become a critical infrastructural node connecting the Italian A34 and Slovenian highways. By facilitating merchandise transport from the High Adriatic port system and Ronchi dei Legionari Airport, it holds the potential to generate significant positive territorial implications for the entire Gorizia region.

References

1. Danielis R (1996) I trasporti e l'ambiente. Aspetti di economia e politica dei trasporti
2. Rothengatter W (1994) Do external benefits compensate for external costs of transport? Transp Res Part A Policy Pract 28:321–328
3. Defrancesco E, Rosato P, Rossetto L et al (2006) Il risarcimento per danno ambientale: aspetti teorici e operativi della valutazione economica
4. Crisman B, Martellos L, Rosato P, Carlo M (2018) La valutazione degli investimenti per la sicurezza delle traverse urbane: un approccio gerarchico multiattributo, pp 1–16
5. Moisello AM (2008) L'activity based costing nelle decisioni d'impresa di breve e lungo periodo. Giuffrè Editore
6. HM Treasury (2018) The green book central government guidance on appraisal and evaluation

Chapter 12
Conclusions

Abstract The research conducted has facilitated the development of an analytical methodology for evaluating the growth prospects of the Gorizia freight terminal. Specifically, this methodology serves as a decision support tool that supplements conventional Multi-Criteria Analysis (MCA) techniques. To this end, the study has enhanced the analysis of governance dynamics in a group decision-making context.

The research conducted has facilitated the development of an analytical methodology for evaluating the growth prospects of the Gorizia freight terminal. Specifically, this methodology serves as a decision support tool that supplements conventional Multi-Criteria Analysis (MCA) techniques. To this end, the study has enhanced the analysis of governance dynamics in a group decision-making context.

As previously discussed, the evaluation model was specifically designed to do more than simply compare alternatives. Instead, it functions as an instrument for understanding the decision-making environment and as a mechanism for balancing various interests. The multi-criteria techniques employed in the analysis -including Simple Additive Weighting (SAW), the simplified Simos method, and the case-based MACTOR method- each contribute differently to identifying strategic recommendations based on representative criteria and parameters of the territorial framework.

This outcome was achieved through a series of iterative evaluations that progressively increased in complexity. Initially, the analysis separately assessed short-, medium- and long-term performance separately, based on the criteria identified by stakeholder, which were categorized as economic or well-being criteria. Subsequently, an approach was developed to enable coherent comparison of the performances of different alternatives occurring at different times and over varying periods. This involved aggregating the obtained values into a single measure, taking into account the duration of the effects over different time horizons.

This methodology provide a deeper understanding of the temporal dimension of decision-making, illustrating how impacts vary over time and must be estimated more precisely. It also enabled challenges associated with balancing long-term strategic vision with immediate needs to be resolved. The results show that incorporating the

E. M. Bertolini, *Temporal Value in MCDM*, Lecture Notes in Intelligent Transportation and Infrastructure, https://doi.org/10.1007/978-3-031-95485-6_12

time factor into evaluations shifts the values of the Alternative/Actor matrices, sometimes leading to significant changes in aggregated outcomes. This finding highlights the importance of incorporating the elicitation of temporal preferences into the MCA model to evaluate their influence on decision-making processes.

Furthermore, the feasibility of implementing any given alternative depends heavily on stakeholder consensus. If agreement among actors is reached, implementation is immediate; conversely, the absence of consensus makes implementation nearly impossible. Given the complexity and lack of transparency that often characterise such intricate, multifaceted systems, incorporating an analysis of potential alliances and conflicts into the evaluation process was invaluablefor accurately representing reality. The study revealed that the interaction between converging and diverging interests is not solely determined by the hierarchical ranking of alternatives for each stakeholder, but also by certain stakeholders' ability to impose their preferences on others. This highlighting the importance of power dynamics.

Thus, it became necessary to analyse power relations through a matrix that records one actor's influence over another. The findings indicate that stakeholders in the examined system possess varying degrees of autonomy, which they can leverage to pursue preferred development trajectories, including forming strategic alliances.

The initial expectations and insights gained throughout the study were validated by the results obtained, as well as by the model's potential for further development. When it was necessary to establish the temporal value of a given alternative, an initial multiperiod aggregation model within the MCA framework had to be constructed — an approach that is not well developed in the existing literature.

Additionally, it was observed that research on participatory MCA remains relatively sparse. To address this, a comprehensive analytical tool was developed to evaluate actor behavior. Without such a tool, MCA would lack the necessary depth to serve as an adequate decision support mechanism.

It is believed that the proposed evaluation model for development prospects could be a valuable support mechanism in territorial governance processes. Furthermore, with the necessary adaptations, it could be applied to other contexts in the future.

In terms of the broader objectives of this research, studying evaluative methodologies applicable to infrastructure transformation scenarios has resulted in the definition of a structured process for transport system planning. Based on qualitative methods, this process enables an in-depth examination of the balance between multiple interests while highlighting power asymmetries among key stakeholders.

The developed tool has proven to be an effective decision support mechanism, as it facilitates a rational approach to addressing the primary challenges that emerge from the territorial scenario. It also fosters greater consensus among all parties involved in the decision-making process.

The analytical component was crucial in addressing the applied evaluation aspect, which is closely related to the subject of assessment — namely, the focus of the planning process. Using the method provided a better understanding of the contribution of decision-analysis frameworks to territorial planning, especially in group decision-making situations, due to the participatory nature of the process. Furthermore, it enabled the quantification of the temporal value of decisions, thereby raising decision-makers' awareness of the short-, medium- and long-term impacts of their choices.

The proposed approach and tool also facilitate the analysis of actor dynamics by quantifying their freedom of action and their political and administrative constraints. Using this user-friendly, spreadsheet-based tool enables alternatives to be assessed in terms of both their technical and political feasibility.

Rcgarding thc spccific casc study, the findings suggest that the Gorizia freight terminal ould operate as an efficient logistics hub, especially when integrated into a wider operational network covering the whole Friuli Venezia Giulia region. Connecting with hinterland and intermodal logistics systems would benefit SDAG and other regional freight terminals by facilitating the development of a new organisational model based on networking and service sharing. To fully leverage expertise and unique capabilities of all supply chain stakeholders in the region, this process must involve them all. This will enable the organisation of integrated logistics services.

SDAG already has a favourable geographical location and benefiting from access to the TEN-T transport corridors. It can accommodate businesses across various industrial sectors and operates state-of-the-art management and warehousing systems that comply with the highest quality standards. Its commitment to sustainability and process optimisation further cements its position as a key driver of economic development in the Isontino area.

However, despite these advantages, questions remain regarding the enhancement of the freight terminal and the development of the regional logistics system. Such improvements would require upgrading the port's infrastructure and strengthening rail and intermodal transport lines, thereby aligning with European strategies to establish the Baltic-Adriatic and Corridor V transport corridors.

MIX
Papier aus verantwortungsvollen Quellen
Paper from responsible sources
FSC® C105338

If you have any concerns about our products,
you can contact us on
ProductSafety@springernature.com

In case Publisher is established outside the EU,
the EU authorized representative is:
Springer Nature Customer Service Center GmbH
Europaplatz 3, 69115 Heidelberg, Germany

Printed by Libri Plureos GmbH
in Hamburg, Germany